POLITICAL MODERNISATION AND THE ENVIRONMENT

ENVIRONMENT & POLICY

VOLUME 24

The titles published in this series are listed at the end of this volume.

Political Modernisation and the Environment

The Renewal of Environmental Policy Arrangements

Edited by

Jan van Tatenhove

Department of Environmental Policy Sciences,
University of Nijmegen,
Nijmegen, The Netherlands
and Department of Political Science,
University of Amsterdam,
Amsterdam, The Netherlands

Bas Arts

Department of Environmental Policy Sciences,
University of Nijmegen,
Nijmegen, The Netherlands

and

Pieter Leroy

Department of Environmental Policy Sciences,
University of Nijmegen,
Nijmegen, The Netherlands

KLUWER ACADEMIC PUBLISHERS
DORDRECHT / BOSTON / LONDON

A C.I.P. Catalogue record for this book is available from the Library of Congress.

ISBN 978-90-481-5459-3

Published by Kluwer Academic Publishers,
P.O. Box 17, 3300 AA Dordrecht, The Netherlands.

Sold and distributed in North, Central and South America
by Kluwer Academic Publishers,
101 Philip Drive, Norwell, MA 02061, U.S.A.

In all other countries, sold and distributed
by Kluwer Academic Publishers,
P.O. Box 322, 3300 AH Dordrecht, The Netherlands.

Printed on acid-free paper

Language editor: Annemarie Weitsel

Contents

Preface

The chapters of this book are a result as well as a representation of the research programme carried out by the Department of Environmental Policy Sciences of Nijmegen University in the Netherlands over the last four years. A core element of this research programme concerns the recent renewal of policies in different domains of environmental politics. Here we discuss our theoretical concepts - political modernisation and policy arrangements - and empirical findings regarding the institutionalisation of environmental politics, both at national and at international level. The empirical evidence has been drawn from a variety of subdomains in environmental politics, including infrastructure policies, nature conservation, climate change, biodiversity and agriculture.

Through this book we hope to engage in further debate with colleagues who also focus on environmental politics, albeit starting from different theoretical concepts or different levels of analysis, such as for instance policy networks, discourse coalitions and advocacy coalitions. Additionally, the book is intended to be used as a textbook in university education programmes on environmental politics.

Jan van Tatenhove
Bas Arts
Pieter Leroy

Introduction

THE EDITORS

Since the 1970s environmental problems have been widely recognised as important social and political issues. The soil contamination of Love Canal (from the 1930s), the contamination from a fertiliser plant in Minamata Bay, Kynshu, Japan (from the 1950s), the fire at the Windscale plutonium production plant (1957), the sinking of the Torrey Canyon (1967), and the chemical fire in the Cuyahuga River (1969) are but a few frightening examples of catastrophes which awoke an environmental awareness among scientists and the public. This awareness not only concerned public health, but also became broader in an ecological sense. The destruction of nature by man affected the 'sustenance base of ecosystems' (Schnaiberg, 1980) and through that the living conditions of human beings on earth.

Growing public attention, the birth, world wide, of the environmental movement and the results of scientific research in the early 1970s, have been paralleled by the gradual institutionalisation of environmental politics. The first reaction of almost all Western nation states to the upsurge in environmental concern was the development of environmental legislation, and the creation of ministries and other administrative bodies to deal with environmental problems. In his analysis of the development of environmental politics in seven countries Weale states: 'in the face of growing political and public concern, virtually the first thing that governments did was to establish a council of independent experts (median date 1970) and a branch of the bureaucratic machine (median date 1970) that had the responsibility for the development of environmental policy' (Weale, 1992: 14). No matter which country, the first steps towards institutionalisation were quite similar.

During the 1980s and 1990s an organisational and substantive renewal of environmental politics has taken place. This renewal was partly in reaction to the drawbacks of environmental politics in the 1970s, which have been labelled 'implementation deficit', 'fragmented institutionalisation' and 'problems of cross-media transfer'. This process of renewal, however, was quite dissimilar in Western European countries such as the Netherlands, Germany and Britain (Boehmer-Christiansen, 1991; Weale, 1992; Hajer, 1997; Jänicke and Weidner, 1995 and Jänicke et al., 1997). In this introduction we will begin by giving a short overview of the convergence, the divergence and the dynamics of environmental politics in these three countries, as an illustration of the diversity of developments in the substance and the organisation of environmental policy.

THE DYNAMICS OF ENVIRONMENTAL POLICY

Since the 1980s British, Dutch and German environmental politics have indeed followed divergent development paths. In general, 'Germany has moved from a position of reluctant environmentalism to one in which it is now legislating some of the most stringent pollution control standards in Europe and pressing internationally for more vigorous action on a wide range of issues. The UK, by contrast, has been laggardly in its adoption of environmental measures, and has acquired the reputation in international negotiations of resisting the development of more forceful pollution control' (Weale, 1992: 72). The Dutch example has seen environmental policy develop into a public management concern, relying largely on the internalisation of environmental goals by target groups. Internationally the Netherlands was regarded as a guide-country, especially at the end of the 1980s and in the early 1990s, when the first National Environmental Policy Plan (NEPP, published in 1989) was considered to be the outstanding example of modern environmental policymaking.

There are some striking differences between the three countries. Firstly there is a difference in principles guiding the process of institutionalisation. Whereas German and to a lesser extent Dutch environmental politics in the 1980s was influenced by the 'precautionary principle' (Vorsorgeprinzip), British environmental politics was strongly influenced by the 'principle of the scientific burden of proof'. The predominant line of the British discourse implies that the scientific understanding of cause-and-effect or doses-impact relationships in natural systems is a necessary precondition for adequate and rational policy making. As a result, technical experts and scientific advisory committees play a crucial role in British environmental policy. This, combined with scientific scepticism about the effectiveness of measures that could be taken, meant that a vigorous development of pollution control was unlikely.

Conversely, the precautionary principle forces policy makers to go 'beyond science' (Weale, 1992: 80), because decisions have to be made in situations where the consequences of alternative policy options are not scientifically determinable within a reasonable margin of error, and where potentially high costs are involved in taking action. Examples are soil contamination and cumulative pollution in industrial areas. In this context politicians are forced to take decisions under conditions of scientific uncertainty, open to public debate and leaving considerable discretion to judicial approbation. The evolution of the precautionary principle in Germany and the Netherlands, however, was quite dissimilar. In Germany this principle was implemented as the 'Stand der Technik' (Best Available Technology) principle, which forced industries to apply the most modern technology and allowed the state to judge them according to general principles and standards. This means that the actual implementation relies upon the expertise of lawyers, engineers and others involved, while at the same time providing opportunities for citizens and the environmental movement alike to appeal to the courts. Over the last few years especially the nuclear issue - still very sensitive in German politics - has given rise to some political and judicial conflicts within German society.

In the Netherlands, the implementation of the 'precautionary principle' was part of the *verinnerlijking* (internalisation) discourse and target group policy. The development of this strategy was a reaction to the implementation deficit and to discussions about government regulation, aiming at deregulation and privatisation. New responsibilities were formulated for involved parties, resulting in a shift from direct regulation to market self-regulation and experiments with economic instruments.

Because different principles guided the institutionalisation of environmental politics, receptiveness to for example ecological modernisation is quite dissimilar in the Netherlands, Germany and the United Kingdom (Weale, 1992; Hajer, 1997; Boehmer-Christiansen, 1991). In general ecological modernisation refers to an ecological transformation of the industrialisation process towards a direction in which the maintenance of the 'sustenance base of society' can be guaranteed, while also reconciling economic growth and the ecological imperative. The German and the Dutch scenarios provided a relatively hospitable environment for ecological modernisation, since their idea of environmental politics largely relied upon new technologies and the conciliation of economy and ecology. British debates, on the other hand, never escaped from the belief that there was inevitable tension between environmental protection and economic development (Weale, 1992: 88; Hajer, 1997). This may be the reason for the quite different design of environmental politics in Britain as far as basic rules are concerned. British politics, as Weale states, in general does not rely upon broad principles, but upon solving a series of problems, each constituting cases that have to be judged on their own merits. The same also applies to environmental politics. Therefore, in contrast to Germany, general norms must be avoided, the particular is preferred to the general, the concrete to the abstract, and unplanned and incremental decision making to medium- and long-term planning.

Another difference is the influence of the environmental movement and the Green Parties in the three countries. In Germany environmentalism was part of a more encompassing opposition movement (Ausserparlementarische Opposition) of (mainly) intellectuals, which emerged in the early 1970s and radically criticised capitalism, science and technology and liberal democracy. Nuclear power and nuclear waste were among the most important issues brought forward by this movement in trying to sensitise and mobilise large parts of German society. However, the environmental movement, while clamouring for political influence, barely succeeded in getting entry to the established political arenas, never mind about getting real political influence on actual decision making, whether on nuclear power, airport enlargement or anything else. This lack of access and influence was caused by and in turn reinforced the political radicalism of the environmental movement. In the late 1970s this radicalisation resulted in the founding of the Green Party, and their successful electoral campaigns forced both traditional parties and the state to develop and promote a more comprehensive environmental policy. Full acceptance by state agencies of environmental issues occurred only at the end of the 1980s, when they began to urge modernisation of environmental politics. Illustrative of this development is the renaming of the environmental department as Ministerium für Umwelt, Naturschutz und Reaktorsicherheit (Ministry for Environment, Nature Conservation and Reactor Safety) after the Chernobyl accident of 1986, but also the coalition government of Social Democrats and Greens after the 1999 elections.

The United Kingdom and the Netherlands experienced much less environmental radicalism than Germany or, for example, France. The British environmental movement, unlike its German counterpart, had a very small 'deep ecology' component, which never had the influence of the German 'fundi's'. Apart from some specific and occasionally violent local and regional environmental conflicts, the environmental issues were only partly politicised in the UK. One of the reasons is the long British tradition of nature conservation as a self-evident preservation of cultural heritage, a tradition which implies well-established policies on nature conservation and countryside preservation. Because these issues and interests are widely accepted, there is no need for politicisation. This situation also entails public and political acknowledgement of the private groups and organisations involved, including their access to political arenas and the acceptance of their lobbying activities in high policy making circles. During the 1980s the environmental movement had very limited influence on the agenda-setting of environmental issues such as acid rain or nuclear power. This was partly as a result of the institutional setting of the British two-party political system, and partly because the political agenda was dominated by economic and 'classical' political issues regarding the competence and responsibilities of state, market and civil society, which concerned the future of the mining industry, the position of the trade unions, the public health service, local government, and the privatisation of a series of national industries and services.

In the Netherlands the environmental movement has a quite different position. In the consensus tradition of Dutch policy making it is frequently consulted about many environmental issues. In fact there is a remarkable relationship between the environmental movement and the Ministry of the Environment. While in the 1970s this relationship was more or less antagonistic, during the 1980s the environmental movement became a natural ally of the Ministry of the Environment. Not only were environmental groups subsidised, they were also invited to contribute to the policy-making process itself. Only the nuclear power debate led to any kind of radicalisation and politicisation of parts of the environmental movement. As a result of the consensus politics, there is less necessity to develop an oppositional strategy in the Netherlands than in Germany and the United Kingdom. In the 1990s the agenda setting of environmental (and other post-material) issues is ensured in negotiations between political parties, the administration, target groups and the environmental movement by a new version of consensus politics: 'the green polder model'.

This short presentation of the institutionalisation of environmental politics reflects the diversity of developments in different countries, both concerning the substance and the organisation of environmental policy. Firstly leading environmental discourses, which structure the substance and the organisation, influence day-to-day policy making. Secondly, the example of the influence of the environmental movement illustrates that the specific relationship between state, civil society and market not only sets the rules for the policy making process in general, but also structures the agencies and the kind of discourses which are legitimate in the given context.

THEME OF THIS BOOK AND CENTRAL QUESTIONS

In the short overview of the dynamics of environmental politics we roughly sketched the theme of this book. We want to analyse and understand the nature and direction of processes of organisational and substantive renewal of environmental politics. Why do the organisation and the substance change? Which external forces influencing these changes can be detected? How does this renewal of environmental politics relate to more comprehensive changes in politics in general? Or is renewal only an internal governmental affair? These topics, however, are not new. Especially in the last two decades, a number of social and political scientists have studied the renewal of environmental politics as a process of institutionalisation. In general, these analyses pay attention to (a) the 'naming' of environmental problems as social problems and their 'framing' as political problems; (b) the sequence of consecutively prevailing styles of governance and the instrumentation of environmental politics and (c) the organisational design of environmental politics. Although there are striking differences between these studies, they predominantly suggest a unilinear evolution in either the naming and framing of environmental problems, or in the prevailing styles of governance, or in the instrumentation and organisation of environmental politics. As regards the styles of governance, for instance, most studies sketch a development 'from an initiating government to a withdrawal-of-the-state' logic, paralleled by a shift 'from regulation to communication strategies'.

Our central line of argument throughout this book is that because in most studies not only a unilinear perspective dominates, but they also have an instrumentalist focus, they do only partial justice to the multiplicity and complexity of developments and transformation in environmental politics. For this reason we will present an analytical framework consisting of the concepts institutionalisation, political modernisation and (environmental) policy arrangements. We believe this framework gives us a more sophisticated tool and enables us to typify and to explain the plurality and diversity of the directions environmental politics has taken. We are especially interested in the interplay between 'traditional' and 'new' policy arrangements in the different stages of the (re-) institutionalisation of environmental politics. In general, institutionalisation of environmental politics results from a series of interactions and interrelations between interdependent agents: politicians, bureaucrats, and representatives of environmental movements, corporate executives and citizens. Their mutual relationships and interactions in the context of more general processes of social and political change shape the processes of environmental policy making in a specific way. This interplay or duality between interaction and macro-processes of social and political change is the central theme of this book. These macro-processes will be understood as political modernisation. In short, political modernisation refers to the change in relationships between the state, the market and civil society in the political domains of (Western) societies, as well as to new concepts of governance. We will argue that those specific processes within environmental policy making and more long-term processes of change converge, resulting in the permanent production, reproduction and renewal of policy arrangements. A policy arrangement is understood as the substantive and organisational stabilisation of a policy domain, consisting of four dimensions: coalitions, rules, resources and discourses. In this sense the concept 'policy arrangement' links long-term processes of political and social change with day-to-day policy making processes.

An important reason for developing the concept of policy arrangements is to enable the institutionalisation of environmental policy making to be understood as the duality between interaction (agency) and social change. By studying the institutionalisation of policy arrangements in environmental policy from the perspective of duality of structure, it is possible to combine two distinguishable but complementary ways of analysis. The first level of analysis (structural social change) focuses on the formation of policy arrangements as a result of changing relations between state, market and civil society. At this level of analysis we emphasise the genesis of the content, the instrumentation and organisation of environmental politics in its broader societal and political context. Secondly, analysis at the level of interaction emphasises the arguments actors use in interaction, the norms and values actors stand for, their problem definitions, and their concepts of the society-nature relationship and societal and political responsibilities. At this level of analysis we look especially at the impact of the arguments and discourses the actors involved produced. In short, we focus upon the impact of the cultural dimensions of interactions, upon the gradual emergence and (re)construction of policy arrangements, and upon the way these policy arrangements are embedded in institutions.

Combining these two levels of analysis provides us with the tools to understand the way agents produce arrangements in day-to-day policy making in environmental policy, at the same time reproducing these arrangements, and the way the environmental policy domain is influenced by more comprehensive processes of change in a macro-political context.

This brings us to the core questions of the book, which have been the central guidelines in writing the chapters to come:

- What kind of environmental policy arrangements have been formed in environmental politics?
- How do these environmental policy arrangements change?
- Is there a shift from one type of environmental policy arrangements to other types? If so, how can these shifts be explained?
- Do different types of environmental policy arrangements co-exist in certain periods of environmental politics? If this is the case, how can this co-existence be explained?
- To what extent are policy domains organised into specific arrangements? In other words, are environmental policy arrangements issue-specific?

THE CONCEPTS OF POLICY ARRANGEMENTS AND POLITICAL MODERNISATION: THEIR POSITION IN THE DEBATE

We will develop the theory of political modernisation and policy arrangements through the discussion of academic debates on the relationship between 'environment' and 'politics' in several disciplines. To expand the concept of policy arrangements we firstly address theoretical notions in politics and policy making, such as policy networks, discourse coalitions and the advocacy coalition framework (Hajer, 1997; Sabatier, 1987; Sabatier and Jenkins-Smith, 1993; Eberg, 1997; Termeer, 1993; Kickert et al., 1997). Secondly we pay attention to the way environmental knowledge, risks and problems are socially constructed and how this influences the processes of agenda setting and the institutionalisation of policy making (Yearley, 1992; Hannigan, 1995; Redclift and Benson, 1994; Dryzek, 1987 and 1997). To understand processes of political modernisation we refer to general debates on change and transformation in sociology and environmental sociology.

At first sight the notion of policy arrangements shows similarities to notions such as policy networks, discourse coalitions, the advocacy coalition framework and the configuration approach. All of these focus on agency, interactions and interdependencies.

In public administration science, policy networks are defined as 'more or less stable patterns of social relations between interdependent actors, which take shape around policy problems and/or policy programmes' (Kickert et al., 1997: 6). Characteristic of the policy network analysis is its focus on the improvement of policy making through concentration on governance and management within constructed policy networks. The goal is to find solutions to policy problems by changing relations of interactions, by formulating new definitions of problems and by introducing new policy instruments. In general, policy network analyses embody attempts to understand and to explain new ways of governance in the context of changing interrelations between state, civil society and market (Benson, 1975; Scharpf et al., 1976; Hanf and Scharpf, 1978; Marin and Mayntz, 1991; Marsh, 1992 and Rhodes; Van Waarden, 1992).

Comparable with the policy network approach is the configuration approach, which sees ongoing interactions between actors as the motor of policy change (In 't Veld et al., 1991). In this approach social and cognitive dimensions are distinguished. Whereas the cognitive dimension is concerned with the construction and the changes of definitions of reality, the social dimension focuses on patterns of interaction and the rules of interaction. The ensemble of actors, definitions of reality, patterns of interaction and rules is called 'social-cognitive configurations' (Termeer, 1993). The binding element of a specific configuration is a shared set of definitions of reality on key issues in a policy field. The constitutive process of a configuration is the 'ongoing interaction among multiple included actors'. In interaction arguments and information are exchanged, leading to changes in definitions of reality and through that to changed patterns of interaction.

A more structure-oriented perspective is Sabatier's conceptual framework of 'advocacy coalitions'. According to the Advocacy Coalition Framework (ACF), policy change is a function of the interaction of competing 'advocacy coalitions' at the level of a policy subsystem, the changes external to the subsystem and the effect of relatively stable system parameters (social structure, rules and resources). Advocacy coalitions are made up of actors from various organisations that share a set of normative and causal beliefs and which often act in concert. A belief system consists of deep core beliefs, near policy core beliefs and secondary aspects, which are arranged in order of decreasing resistance to change. Policy change within a subsystem is understood as the product of two processes. First policy-oriented learning, which is defined as 'relatively enduring alterations of thought or behavioral intentions that result from experience and are concerned with the attainment (or revision) of public policy' (Sabatier, 1987); (Eberg, 1997). 'Advocacy coalitions within the subsystem attempt to translate the policy cores and secondary aspects of their belief systems into policy programmes' (Eberg, 1997: 21). Second, external perturbation, that is, 'the effects of system-wide events on the resources and constrains of subsystem actors' (Eberg, 1997: 21). These external effects in particular lead to changes in the core aspects of a policy.

Central in Hajer's argumentative approach are discourses, story lines and discourse coalitions. His approach focuses on the constitutive role of discourses in political processes and allocates a central role to the discoursing subjects, in the context of the idea of duality of structure (Hajer, 1997: 58). Politics is seen as a struggle for discursive hegemony in which actors try to secure support for their definition of reality. In his argumentative analysis Hajer introduces two middle-range concepts: story-lines and discourse-coalitions. Story-lines are defined as 'narratives on social reality through which elements from many different domains are combined and that provide actors with a set of symbolic references that suggest a common understanding' (Hajer, 1997: 62). Story lines are supposed to be the essential elements in the clustering of knowledge, the positioning of actors, and, ultimately, in the creation of coalitions amongst the actors within a given domain (Hajer, 1997: 63). 'Discourse-coalitions are defined as the ensemble of (1) a set of story-lines; (2) the actors who utter these story-lines; and (3) the practices in which this discursive activity is based. Story-lines are here seen as the discursive cement that keeps a discourse-coalition together' (Hajer, 1997: 65).

When we compare these approaches (for an extensive comparison see Pestman and Broekhans, 1998) to the study of policy making, the policy network and the configuration approaches are mainly actor-oriented and voluntaristic in nature, and tend to neglect structural, institutional and power features of policy making and implementation. The advocacy coalition model, on the other hand, is more static and structuralistic. In the Advocacy Coalition Framework the institutional context is largely external to the subsystem, and the relationship between subsystems and the institutional context (external parameters and events) is not very clear. The picture that emerges from this is one of a rather autonomous structure, determining the actions of actors (Pestman and Broekhans, 1998). Although the policy arrangements approach incorporates elements of these methods of studying policy making, it bears most resemblance to Hajer's argumentative approach: in both the duality of structure is the starting point of analysis. In his concept of discourse Hajer incorporates the institutional dimension. Following Foucault and Giddens, the main theoretical thesis of the argumentative approach is 'that one can observe how the institutional practices in the environmental domain work accordingly to identifiable policy-discourses that through their story-lines provide the signpost for action within these institutional practices' (Hajer, 1997: 264). In his critique of Sabatier Hajer emphasises his relational ontology, the constitutive role of language and the role of story lines and discursive affinities, and his theory of social change (Hajer, 1997: 68-72). Story-lines, determining the interplay between physical and social realities, are seen as the vehicles for change and are analysed in connection with the specific discursive practices in which they are produced (Hajer, 1997: 72).

In contrast to discourse-coalitions, the focus in policy arrangements analysis is not so much on argumentative and discursive activity within a policy arrangement, but more on the institutionalisation of an arrangement in the interplay between discursive activity and processes of social and institutional change. We have therefore developed the concept of political modernisation, inspired by authors such as Beck, Giddens, Kumar and Castells. On the one hand, the environmental crisis is supposed to be a trigger for more general and political transformations, which induce the transition from simple modernity to high or reflexive modernity (Giddens, 1990, 1991 and 1994; Beck, 1986 and 1994; Lash et al., 1996). In the theory of ecological modernisation (Mol, 1995; Hajer, 1997; Spaargaren, 1997) this thesis is refined into a specific theory of political modernisation, with the focus on new forms of political intervention. On the other hand, processes of political modernisation single out the structural and institutional forms and dimensions of policy making on the crossroads of state, civil society and market in a globalising world. In our elaboration on political modernisation we develop a twofold argument, referring to different debates on social and political change. First we distinguish three phases of political modernisation: 'early', 'anti', and 'late'. Each of these phases can be characterised by specific discourses on progress and development (Wagner, 1994), types of governance and domination, and concepts of technology. Secondly we elaborate on relationships between state, civil society and market. This discussion is based on theories about the state and civil society and their interrelations with the market. As stated above, the combined analysis of processes of political modernisation and specific processes of environmental policy making results in the production and reproduction of policy arrangements.

OUTLINE OF THIS BOOK

The book consists of three parts: an analytical framework, empirical results and conclusions.

Part I introduces the theoretical concepts - institutionalisation, political modernisation and environmental policy arrangements. In chapter 1 the development and renewal of environmental politics are analysed as processes of institutionalisation. On the basis of discussions on institutionalisation we conclude that most studies are too unilinear to make the multiplicity and complexity of contemporary environmental politics comprehensible. Therefore we introduce the concepts of political modernisation (chapter 2) and (environmental) policy arrangements (chapter 3). Policy arrangements, as stated above, are the outcome of interactions and interdependencies among actors in environmental policy subdomains, in the context of processes of structural change (political modernisation. This notion of the duality of institutionalisation of policy makes it possible to grasp the plurality and dynamics of environmental politics.

In part II we change the focus to the dynamics and specific institutionalisation of policy arrangements in five (empirical) environmental domains. In chapter 4 Dutch infrastructure policy is discussed. Recently the Dutch government has initiated several large infrastructure projects, such as railways and airports. The decision-making processes surrounding these projects, however, vary. Some are characterised by open arrangements, in which public participation is guaranteed. Others, though, show quite the opposite tendency: top-down decision-making is increased and public participation is reduced. Hence, the process of political modernisation does not show an unequivocal 'face' in the infrastructure policy domain, as two types of arrangements (traditional public participation and experiments with new arrangements) and logic interfere. Additionally, chapter 4 addresses the historical background of the arrangements in Dutch infrastructure policies and focuses on the (changing) role of the several players within this domain: e.g. state departments, citizens' groups and (privatised) companies.

Chapter 5 deals with transboundary policy arrangements in German-Dutch nature conservation policy. Due to, among other things, European integration, the devolution of national policies within individual EU member states, and the international nature of conservation issues, transboundary projects on nature conservation policies have recently emerged at the subnational level. One example is the *Gelderse Poort* project of the *Provincie Gelderland* (Netherlands) and the *Kreis Kleve* (Germany). So far it has become obvious that the current arrangements developed on the two sides of the border will not easily fit together, either in terms of substance or of organisation. After all, the processes of political modernisation were not the same in the two countries, the roles of several players such as the state and conservation organisations differed, and the discourses on nature conservation were not identical in the Netherlands and Germany or in the border regions involved. However, new transboundary networks have emerged. This dynamic of old and new, as well as national and transboundary arrangements, in the nature conservation domain is analysed. The question addressed in chapter 5 is whether such an initiative will lead to a new type of policy arrangements in nature conservation.

Chapter 6 presents an overview of policy arrangements in the global arena. Since 1945 the United Nations (UN) has addressed environmental concerns, intermittently before 1970, but continuously from the 1970s onwards. This intensification of concern was accompanied by changes in environmental discourses and the organisational settings within which global policies were designed and implemented. Briefly, ever more dimensions of environmental issues and ever more players were added or became involved in global policy arrangements as time went on, changing both discourses and organisational settings. In this chapter three periods are distinguished and analysed: 1945-1972, 1972-1982, and from 1982 to the present. The chapter argues that each era coincides with one specific and predominant environmental policy arrangement, although some interfaces between new and old arrangements continue to exist in each phase. In addition, it is argued that in global environmental politics the roles of the nation state and of international organisations, as well as of non-state players, have fundamentally changed. The analysis will be illustrated by examples of global issues such as natural resources, nature conservation, biodiversity and climate change.

Central in chapter 7 is the institutionalisation of green planning. In order to attune the planning systems in the environmental, spatial and water management policy domains, integration of plans has recently been the aim in the Netherlands and other Western countries. Since its emergence in the early 1970s, environmental policy has gradually changed from a more or less ad hoc policy, taking measures in response to specific problems, to a more coherent and planned policy field. Green planning is a widely accepted concept nowadays. Nevertheless, the images behind the concept, the motives for it and, quite obviously, the scope, status and procedures of planning documents, do differ both in time and space. Chapter 7 diagnoses, from an international perspective, the role green planning - as a policy arrangement - has played and still plays in the Netherlands. It depicts its internal programming function from the early stages of environmental policy development, its widening scope and broadening function since the mid-1980s, and its functioning as a policy integration instrument in the 1990s.

Finally, in chapter 8 policy arrangements in rural areas will be discussed. Rural regions in post-modern societies currently experience a radical transformation in economic, political, social and cultural respects. After all, the agricultural sector has to cope with environmental, nature conservation, and animal rights and well-being considerations nowadays, while at the same time it has to survive in ever more competitive and globalising markets. Therefore the neo-corporatist policy arrangements, in which small elites have determined the course of the agricultural sector in the Netherlands, are no longer sufficient. To experiment with alternative rural development, new regional projects have been implemented. These are characterised by new policy arrangements, which are more decentralised, egalitarian and democratic in nature. Although some remnants of old arrangements still exist, it is unlikely that the rise of civil society in contemporary rural policy making will be reversed.

In Part III (chapter 9) conclusions are drawn with regard to the diversity and multiplicity of traditional and new policy arrangements in environmental policy. On the basis of the empirical data in part II we give answers to the central questions of part I. In particular we discuss and explain why different environmental policy arrangements co-exist in certain eras of environmental politics, and why there are disparities between different domains of environmental policy making. Finally, a research agenda will be presented.

REFERENCES

Beck, U. (1986), *Risikogesellschaft. Auf dem Weg in eine andere Moderne*, Frankfurt am Main: Suhrkamp.

Beck, U. (1994), The Reinvention of Politics: Towards a Theory of Reflexive Modernization, U. Beck, A. Giddens and S. Lash, eds., *Reflexive Modernization. Politics, tradition and aesthetics in the modern social order*, Oxford: Polity Press, pp.1-55.

Benson, J.K. (1975), The Interorganizational Network as a Political Economy, *Administrative Science Quaterly*, 20(2), pp. 229-249.

Boehmer-Christiansen, S. and J. Skea (1991), *Acid Politics. Environmental and Energy Policies in Britain and Germany*, London, New York: Belhaven Press.

Dryzek, J.S. (1987), *Rational Ecology. Environment and Political Economy*, Oxford, New York: Basil Blackwell Inc.

Dryzek, J.S. (1997), *The Politics of the Earth. Environmental Discourses*, Oxford: Oxford University Press.

Eberg, J. (1997), *Waste Policy and Learning. Policy Dynamics of Waste Management and Waste Incineration in the Netherlands and Bavaria*, Delft: Eburon.

Giddens, A. (1990), *The Consequences of Modernity*, Oxford: Polity Press.

Giddens, A. (1991), *Modernity and Self-Identity. Self and Society in the Late Modern Age*, Cambridge: Polity Press.

Giddens, A. (1994), Living in a Post-Traditional Society, U. Beck, A. Giddens and S. Lash, eds., *Reflexive Modernization. Politics, tradition and aesthetics in the modern social order*. Oxford: Polity Press, pp. 56-109.

Hajer, M. (1997), *The Politics of Environmental Discourse. Ecological Modernization and the Policy Process*, Oxford, New York: Oxford University Press, Clarendon Press.

Hanf, K. and F.W. Scharpf, eds. (1978), *Interorganizational Policy Making. Limits to Coordination and Central Control*, London.

Hannigan, J. A. (1995), *Environmental Sociology. A social constructionist perspective*, London and New York: Routledge.

In 't Veld, R.J., L. Schaap, C.J.A.M. Termeer and M.J.W. van Twist eds. (1991), *Autopoiesis and configuration theory: new approaches to societal steering*, Dordrecht: Kluwer Academic Publishers.

Jänicke, M. and H. Weidner, eds. (1995), *Successful Environmental Policy. A Critical Evaluation of 24 Cases*, Berlin: Edition Sigma.

Jänicke, M, M. Binder, and H. Monch, (1997), Dirty Industries': Patterns of Change in Industrial Countries, *Environmental and Resource Economics*, 9, pp. 467-491

Kickert, W.J.M., E.H. Klijn and J.F.M. Koppenjan, eds. (1997), *Managing Complex Networks. Strategies for the Public Sector*, London: SAGE.

Lash, S., B. Szerszynski and B. Wynne eds. (1996), *Risk, Environment and Modernity. Towards a New Ecology*, London: SAGE.

Marin, B. and R. Mayntz, eds. (1991), *Policy Networks. Empirical Evidence and Theoretical Considerations*, Frankfurt am Main, Boulder: Campus Verlag, Westview Press.

Marsh, D. and R. Rhodes, eds. (1992), *Policy Networks in British Government*, Oxford.

Mol, A. P. J. (1995), *The Refinement of Production. Ecological modernization theory and the chemical industry*, Utrecht: Van Arkel.

Pestman, P. and B. Broekhans (1998), *An assessment of three social-cognitive models used for environmental public policy analysis*, SISWO, Sociaalwetenschappelijke Studiedagen, Amsterdam.

Redclift, M. and T. Benson, eds. (1994), *Social Theory and the Global Environment*, Global Environmental Change Series, London, New York: Routledge.

Sabatier, P.A. (1987), Knowledge, Policy-Oriented Learning, and Policy Change, and advocacy coalition framework, *Knowledge: Creation, Diffusion, Utilization*, 8(4), pp. 649-692.

Sabatier, P.A. and H.C. Jenkins-Smith, eds. (1993), *Policy change and learning: An advocacy coalition approach*, Boulder, CO: Westview Press.

Scharpf, F.W., B. Reissert and F. Schnabel (1976), *Politikverflechtung: Theorie und Empirie des kooperativen Feuderalismus in der Bundesrepublik*, Kronberg: Scriptor.

Schnaiberg, A. (1980), *The Environment from Surplus to Scarcity*, New York/Oxford: Oxford University Press.

Spaargaren, G. (1997), *The Ecological Modernization of Production and Consumption. Essays in Environmental Sociology*, Wageningen: WAU.

Termeer, C.J.A.M. (1993), *Dynamiek en inertie rondom mestbeleid: een studie naar veranderingsprocessen in het varkenshouderijnetwerk*, Den Haag: VUGA.

Van Waarden, F. (1992), Dimensions and types of policy networks, *European Journal of Political Research*, 21(1-2), pp. 29-52.

Wagner, P. (1994), *A Sociology of Modernity. Liberty and Discipline*, London and New York: Routledge.

Weale, A. (1992), *The New Politics of Pollution*, Manchester and New York: Manchester University Press.

Yearley, S. (1992), *The Green Case. A sociology of environmental issues, arguments and politics*, London, New York: Routledge.

1. The Institutionalisation of Environmental Politics

JAN VAN TATENHOVE AND PIETER LEROY

1.1 INTRODUCTION

In this chapter we focus on the institutionalisation of environmental politics as this domain emerged in Western European countries from the early 1970s. Institutionalisation is used here in its sociological meaning, referring to the construction and the preservation of day-to-day activities and interactions of actors in institutions, within a context of processes of societal and political change. More specifically, institutionalisation is regarded as the process leading to the formation, deformation and reformation of policy arrangements.

After defining the concept in section 1.2 we give a 'state-of-the-art' account of the way social and political scholars have analysed the development and institutionalisation of environmental politics (section 1.3). This analysis focuses on (a) the 'naming' of environmental problems as social problems and their 'framing' as political problems; (b) the sequence of consecutively prevailing styles of governance and the instrumentation of environmental politics and (c) the organisational design of environmental politics. Although there are striking differences between the various studies, they predominantly use an evolutionary perspective. Environmental politics is supposed to be developing from a policy domain initiated and dominated by governmental regulation to 'withdrawal-of-the-state' logic and communication strategies. This portrayal, though plausible on a general level of analysis, does no justice to the actual multiplicity and complexity of developments in environmental politics. That plurality will be studied in detail in part II, gathering and comparing evidence from different cases and subdomains of environmental politics.

We conclude that, although consecutive phases of the institutionalisation of environmental politics can be characterised by predominant styles of governance and therefore by certain types of arrangements, there is not a single leading principle guiding these developments. Within a context of presumed univocal political processes, any nation state and any environmental policy subdomain has its specific combination of dominant actors, discourses, rules and resources, resulting in more or less specific arrangements.

1.2 INSTITUTIONALISATION: THE CONCEPT

Institutionalisation is one of the central concepts used in social and political sciences to study the ongoing processes of transformation by which institutions are (re)produced in interactions. In general institutionalisation refers to the ongoing process of the patterning, preservation, construction, organisation and deconstruction of day-to-day activities and interactions in institutions. As has been described time and again in the social sciences, institutions tend to function as semi-autonomous social agents. The same applies to politics, its institutionalisation processes and its institutions. Wells (1996) emphasises the functioning of political institutions as semi-autonomous actors. Political institutions distribute political resources in a specific way, they determine or modify individual preferences through control of a wide range of symbols, myths and rituals, they define the rules of the game and they impose order in a policy area. Wells' analysis expresses the recent theoretical and empirical resurgence of institutions and institutionalisation in the political sciences, which has become known as 'neo-institutionalism'. In this view institutions indeed play a semi-autonomous role in the political system, as political agencies in their own right. 'Social, political and economic institutions have become larger, considerably more complex and resourceful, and prima facie more important to collective life' (March and Olson, 1984: 734). This 'new' institutionalism emphasises the relative autonomy of political institutions, the possibilities for inefficiency in history, and the importance of symbolic action to an understanding of politics (March and Olson, 1984).

Keman (in Steunenberg et al., 1996) distinguishes three approaches within neo-institutionalism. Firstly the *historical-traditional approach*, which mainly uses a macro-perspective and analyses long-term processes. This approach, therefore, not only assumes that institutions exert influence on policy-making processes, but sees institutions themselves as the result of long-term historical processes. Whereas recently-predominant approaches in the political sciences tend to under-appreciate or even to neglect the role of classical institutions, including the nation state, one result of neo-institutionalism is that 'the state is brought back in'. Keman secondly identifies a *cultural-organisational approach*, aimed mainly at the empirical analysis of processes of institutionalisation at the meso-level. This approach focuses on the relationship between individual perceptions and motivations of 'stakeholders' within and around organisations and the socio-cultural characteristics of the context. The varying institutionalisation of environmental politics - for instance in terms of its delineation from other policy domains, or its organisation - in relation to the specific cultural and organisational characteristics of different countries fits very well into such an approach. Thirdly Keman distinguishes the *(political-economic) rational choice approach.* In this approach, again, institutions are regarded as exerting influence on actual processes of decision making. But the emphasis here is on their influence, both constraining and enabling, upon 'individual' conduct, 'individual' of course including individuals as well as organisations. The focus thus is on the actual influence of institutions and institutional arrangements, such as forms of privatisation, on the processes of individual and collective choices and their eventual outcome.

In this chapter we focus on some historical and cultural-organisational approaches to the institutionalisation of environmental policy, and more specifically its institutionalisation into a wide variety of policy arrangements. In general, institutionalisation is a process of transformation in which policy arrangements are produced and reproduced in interaction, within the context of long-term processes of societal and political change. As stated in the introduction, we analyse processes of institutionalisation from two distinguishable but complementary perspectives. The first perspective pays attention to the formation of policy arrangements as a result of changing relations between state, market and civil society. At this level of analysis we emphasise the genesis of the content, and the organisation of environmental politics in its broader societal and political context. The second perspective emphasises the norms, arguments and values which actors use in interaction. In this kind of analysis the accent is on problem definitions, concepts of society-nature relationships and notions of societal and political responsibilities. More specifically, we look at the impact of arguments and discourses used by actors upon the gradual emergence and reconstruction of arrangements. These two perspectives provide us with the tools to understand the process of institutionalisation as being the interplay between agency and structure.

More particularly, we define institutionalisation as the interplay between structuration and stabilisation, in which the contents and the organisation of policy arrangements are (re)produced in interaction, within the context of long-term processes of societal and political change (Van Tatenhove, 1993). Structuration refers to the (re)production of content and organisation of a policy domain in interaction, whereas stabilisation refers to the 'preservation' of contents and organisation in specific policy concepts and arrangements. Essentially, the process of structuration is about the gradual formation of an arrangement in interactions. The specific patterns of interaction result in more or less commonly accepted rules of the game, coalitions, discourses and problem definitions, acknowledged and legitimised in the context of that arrangement. The process of stabilisation, on the other hand, refers to the situation where arrangements have already been formed, constraining the actors involved into adopting certain concepts, values, patterns of interaction etc. The analytical distinction between structuration and stabilisation should not be considered a dualism, but is understood as a duality of structure. Agents are never completely constrained by (stabilised) institutions or arrangements, nor are they completely free to form or change (to structure) institutions or arrangements. The interplay between structuration and stabilisation refers to the ongoing process of interactions made possible within a context of arrangements, whereas arrangements at the same time are formed and modified by interactions. In the next section we will concentrate on how the institutionalisation of environmental politics is scrutinised in different studies.

1.3 INSTITUTIONALISATION OF ENVIRONMENTAL POLITICS: THE STATE OF THE ART

In this section we present a general impression of the way different authors deal with the emergence of environmental politics and its gradual development over the last three decades. In their analyses most of these authors implicitly use an institutionalisation frame of reference. They often start with the emergence of environmental problems as relatively new social problems in the late 1960s and their societal and political recognition and acknowledgement in the early 1970s, resulting in the gradual development of a new political (sub)domain, with its specific structure, types of governance, organisation and instrumentation. Since new environmental problems emerged in the 1980s (acid rain, global warming and others) while others were redefined, and since the political macro-context changed, the institutionalisation of environmental politics has to be regarded as a perpetual process of construction, deconstruction and reconstruction.

1.3.1 The social and political construction of environmental problems

A first similarity in studies of the institutionalisation of environmental policies is their focus on the social construction and definition of environmental problems.

Though the social constructivist perspective has only recently been fully applied to environmental problems and politics (Hannigan, 1995), the social construction of environmental problems has been studied by environmental sociologists and political scientists from the early 1970s (cf. Downs, 1972; Albrecht, 1975; Schoenfeld et al., 1979). In general this perspective on environmental problems and politics involves two distinguishable but obviously interdependent processes: the definition of environmental problems as social problems on the one hand, and the (re)formulation of these problems into politics on the other. We refer to these processes as respectively societal and political agenda setting. Starting points of such analyses are the definition of environmental problems, the strategies used by interest groups, environmental organisations, professionals and others in naming and framing environmental problems, the changes within these definitions and the underlying discourses over time, and the process of societal and political agenda setting. Other points of interest are the cultural and structural openness of a political system towards newly defined problems (Potter, 1995; Blowers and Leroy, 1996) and the way in which the social and political definition of an environmental problem affects its political handling and its organisational and institutional embedding. These and related questions on the 'naming' and 'framing' of environmental problems into politics have been studied from different theoretical perspectives. If expressed as a continuum, the end points are respectively rational synoptic and social constructivist approaches.

For quite a long time the rational synoptic perspective on agenda setting was dominant in the social and political sciences in general, and therefore it inspired the research into environmental agenda setting in particular. The 'issue-attention cycle' described by Downs (1972) and the theory of agenda building put forward by Cobb and Elder (1972) are classics in this field of study. The 'issue-attention cycle' deals primarily with public attention paid to problems. The cycle itself has five stages: 'the pre-problem stage', 'alarmed discovery and euphoric enthusiasm', 'realising the cost of significant progress', 'gradual decline of intense public interest' and 'the post-problem stage'. These stages may vary in length depending upon the particular issues involved, but they always occur in this sequence. According to Downs (1972: 41) 'problems that have gone through the cycle almost always receive a higher average level of attention, public effort, and general concern than those still in the pre-discovery stage'. Public opinion polls (Albrecht, 1975) and media coverage (Schoenfeld, 1979) have been studied thoroughly for empirical evidence which indicates the stage environmental concern or a specific environmental issue can be classified as having reached. As is clear from the above, this approach works primarily on the assumption that the amount of public attention is an indicator showing when a problem achieves the next stage of the cycle. Other authors therefore have studied the 'attention mobilisation' strategies of both environmental movements and the media. Cobb and Elder (1972) address the same issue from an analytical perspective, by distinguishing three ways to get issues onto the formal political agenda. Firstly, issues outside government are put on public agendas, via which they ultimately gain political attention (the 'outside model'). Secondly, public agents define issues for the formal agenda (the 'inside initiative model') and thirdly, interest groups and policy makers try to insert an issue into the public agenda, hoping that it will reach formal agenda status (the 'mobilisation model'). A lot of studies from the 1970s and early 1980s deal with these questions and provide empirical evidence. The emphasis in these studies is on the mobilisation strategies of the early environmental movement, the role of the mass media and the adoption or rejection of environmental issues on local, national and international agendas. The analysis of the process of agenda building is analytically refined and expanded by Bachrach and Baratz (1970), who introduce the concept of non-decisions, and by Van der Eijk and Kok (1975), who add and analyse different 'barriers' in the successive stages of the Eastonian model of policy making.

Though starting from a quite different point of view, the recently developed social constructionist approach to the agenda-setting process eventually deals with similar questions to the ones listed above. Hannigan (1995), for example, identifies six factors which he judges to be necessary for the 'construction' of an environmental problem. These are scientific authority for and validation of claims; the existence of 'popularisers' who can bridge environmentalism and science; media attention in which the problem is 'framed' as novel and important; dramatisation of the problem in symbolic and visual terms; economic incentives for taking action; and the emergence of an institutional sponsor who can ensure both legitimacy and continuity. Like the rational synoptic approach, the social constructionist perspective explains the upsurge in public attention to and interest in environmental problems in the 1960s through a combination of factors. Environmental catastrophes such as the sinking of the Torrey Canyon (1967) or the chemical fire in the Cuyahuga River (1969) were important, but they were insufficient as triggers for the institutionalisation of environmental politics. Of more importance was the 'framing' of environmental problems in newly developed ecological and scientific frameworks, provided for instance by the first report to the Club of Rome (Meadows, 1972). And like their colleagues using the rational synoptic approach, the authors inspired by the social constructivist perspective emphasise the importance of the mobilisation of both mass media and public attention from the early 1970s (Yearley, 1992; Hannigan, 1995; Liberatore, 1995). The green case, Yearley states, was put forward and the green agenda was shaped by a coalition of interdependent actors. Environmental groups were the first to be engaged in making social problem claims. Through sophisticated ways of campaigning they built up a public profile of environmental problems; news items were delivered, celebrities were brought into action and scientists provided the scientific knowledge in claiming that these were social problem.

The similarities in questions and conclusions between the two perspectives, however, should not disguise the differences. Whereas the rational synoptic approach is primarily devoted to quantitative increases in attention and support, caused by mobilisation strategies aimed at overcoming the subsequent barriers on the road to the political agenda, the social constructionist approach pays much more attention to the qualitative aspects of the 'naming' and 'framing' of environmental problems. These aspects include the shifts in what Buttel et al. (1990) still labelled 'environmental ideology', and what has been called 'environmental discourse' since. When studying the role of environmental movements and their actual influence upon agenda-setting processes therefore, the rational synoptic approach ends up with an analysis of the 'resources' these organisations have succeeded in mobilising. An example is the resources mobilisation approach elaborated by Zald and McCarthy (1979) and used by many authors (Bader, 1991; Fisher and Kling, 1993). Authors inspired by social constructivism, on the other hand, will and do emphasise the cognitive and argumentative framing that the environmental movement (or other parties involved) use to encourage environmental concern and attention. Therefore authors such as Buttel et al. (1990), Buttel and Taylor, (1992), Yearley, (1992), Hannigan, (1995), Liberatore, (1995), Hajer, (1997) and many others pay particular attention to the arguments and discourses behind (un)successful agenda setting. Hajer (1997) for instance tries to find out why and how the 'ecological modernisation' thesis could have been so successfully promoted as the key concept of contemporary environmental politics. This approach, apart from its social constructivist and post-modernist roots, has largely been inspired by the argumentative and discursive approaches (social learning, socio-cognitive configurations, advocacy coalitions and others) that mainly seem to dominate the political sciences of the 1990s.

In short, the theme of the social and political construction, the 'naming' and 'framing' of environmental problems, has been extensively commented upon by social and political scientists. Above we distinguished what might be the two extremes of a continuum. Firstly, the application of a rationalistic or synoptic approach to agenda setting and agenda building on environmental issues, focusing on how these issues are inserted into public and political agendas. Secondly, the application of argumentative approaches of social construction to environmental issues, focusing on how these issues have been formulated in order to deal with them politically. The vast majority of authors and approaches of course lie between these presumed extreme rationalistic and extreme social constructivist points of view. As mentioned above, the 'issue-attention' cycle and the agenda-building and barrier models have been complemented and sophisticated over time by more interactive or constructivist models, varying from 'garbage-can models' (Cohen, 1972; Kingdon, 1984) to discourse coalitions (Hajer, 1997).

1.3.2 The instrumentation of environmental policies

The second issue many authors have dealt with concerns the (changing) form of governance within the environmental policy domain, especially as a changing governance style is reflected by the shifting instrumentation of environmental politics. With regard to this issue, there is largely a consensus among the authors (cf. Weale, 1992; Hajer, 1997; Mol, 1995; Hanf, 1996; Garner, 1996), as they all report a twofold gradual shift. First, the instrumentation of environmental policies has shifted from regulatory strategies via market instruments towards more communicative and interactive policy strategies. Secondly, these strategies evolved from narrowly defined, highly fragmented sectoral regulation, via more comprehensive instruments (environmental impact assessment and the like), towards strategies which address the driving forces of environmental pressure, such as agriculture, industry, traffic etc. and which aim to modernise these activities towards more sustainable development. This second change is directly related to the changing social perception and political definition of environmental problems, in other words to its varying 'naming' and 'framing'.

The change 'from regulation via market to communication' applies to both (Western!) nation states and supranational organisations, e.g. EU policies, whereas the development of international regimes and strategies reveals a much more varied pattern (see chapter 6). Weale's analysis (1992) is highly representative of the writing of other authors. When in the late 1960s and early 1970s environmental problems were put on the agenda, there was great optimism about the problem-solving ability of technical measures on the one hand, and of state-initiated regulatory intervention on the other. The first led to the introduction of end-of-pipe technologies to solve the most threatening environmental problems, such as water and air pollution, noise pollution and waste production. The introduction and general use of these end-of-pipe technologies (such as purification, sewage treatment, the building of higher chimney stacks, the diffusion of waste water etc.) was implemented and reinforced through a system of generalised emission and immission standards, put into operation by means of individual permits. This meant that states, e.g. governmental bodies, reverted to a traditional set of instruments, which had been institutionalised in one form or another in almost all Western European nation states since the Napoleonic laws on hazardous activities. During the nineteenth century these regulations were expanded to include working conditions and labour safety measures in most countries, but until the 1970s governmental concern with external nuisance was rather limited. Yet they formed the basic model for state-initiated environmental regulations in the early 1970s.

Within a few years, however, these regulations were transformed into specific laws and standards concerning the different spheres of the environment: water, air, waste etc. As a result, the first stage of environmental policy in almost every Western country consists primarily of a list of laws and standards. Examples are the 1970 Clean Air Act in the United States; the 1974 Federal Immission Control Act in West Germany; the 1974 Control of Pollution Act in the United Kingdom; the 1961 Air Pollution Law and the 1964 Water Law in France; the 1969 Surface Water Pollution Act and the 1970 Air Pollution Act in the Netherlands. At the end of the 1970s, however, the drawbacks of this regulatory strategy and its sectoral fragmentation became obvious, mainly through three interrelated problems. The first was the 'implementation deficit'. The question of implementation was put forward at first by Pressman and Wildawsky (1973), while Mayntz, (1978) described the implementation problems contained in German environmental policies. Empirical evidence from Germany, the Netherlands and other countries made clear that the actual implementation of the regulatory policy system was highly problematic, as large parts of the intended policy (in terms of standards and permits to be applied, checks and controls to be organised) was never put into practice. We cannot go into the reasons and details here, but the implementation deficit clearly illustrated the limited abilities of governmental bodies to govern-by-regulation. In the 1980s this evidence was put forward as an argument for the withdrawal of the state, for deregulation and for the reinforcement of environmental policies through market mechanisms.

Two other problems became manifest in the same period: the 'fragmented institutionalisation' of environmental policy (Van Tatenhove, 1993) and the problems of cross-media transfer. Fragmented institutionalisation refers to the fact that the competency for the designing and implementation of environmental policy was divided among a number of ministries or administrative bodies, all dealing with different aspects of the environment. That fragmentation, firstly, in itself complicated the implementation problem, since every single citizen or entrepreneur applying for a permit had to deal with very different governmental bodies, all looking separately at one single aspect of the issue, thereby multiplying the workload of both governmental agencies and their clients. Secondly, this approach caused the problem of cross-media transfer, because the policy was intended to protect only a particular environmental medium, such as air, thereby neglecting the fact that regulation of air pollution could easily create landfill or water pollution problems (Weale, 1992). As polluters always choose the line of least resistance, the fragmented institutionalisation of environmental policy resulted in a lack of ecological integration.

The implementation deficit and the problems of fragmented institutionalisation and cross-media transfer first of all led to efforts in almost all Western countries to simplify and streamline the environmental regulatory system. Linking the different standards and combining the different permit systems, in short, *co-ordination and integration*, were the catchwords of the new strategy. In the Netherlands, the introduction of the environmental impact assessment - in which one was obliged to look at all possible environmental impacts at the same time - played a crucial role in the renewal of environmental legislation from the late 1970s onwards. The second strategy aimed to overcome the fragmented and *ad hoc* character of separate laws, regulations and permits by designing a more comprehensive and long-term environmental policy. This led to *green* or *environmental planning*, the background, aims and realisation of which will be described in chapter 7. Both strategies implied a renewal of the instrumentation of environmental policies, although both integration and planning still presume a state-initiated style of governance.

But secondly these and other efforts to overcome the drawbacks in the institutionalisation and implementation of environmental policies were influenced by more general political discussions about the role of the state from the late 1970s onwards. As a result of the economic crisis in the 1970s, the idea that the state could (or even should try to) blueprint societal developments gradually disappeared. The belief in planning and detailed regulations came under attack, leading to political discussions about the scope and the level of detail and complexity of governmental regulation. As is well known, these discussions eventually dealt with the changing role of the state *vis à vis* the market and civil society, that is, with changing interrelations and responsibilities, to which we will return below. Although both discussions are interdependent, we will initially restrict ourselves to the instrumentation aspect of environmental policy.

Though the idea of economic instrumentation of environmental policies has been mooted since the early 1970s (Baumol and Oates, 1975; and many others), the 1980s saw growing interest in a non-regulatory approach to the environmental policy of Western societies, especially in economic instruments (Opschoor and Vos, 1989). This plea for economic instruments was not only inspired by the drawbacks of the regulatory strategy described above. It was at the same time an attempt to bring into practice one of the long-lasting principles of environmental policy, namely 'the polluter pays' (the normative argument). In addition, economic instruments can be expected to have a more effective impact on target groups, because taxes and subsidies influence the weighing up of behavioural alternatives by appealing to self-interest (the functional argument). Therefore, both at national and at international level experiments with a wide variety of economic instruments were introduced. Taxes, levies and subsidies were established and rapidly institutionalised on both the national and the supranational (e.g. EU) level, while the bubble concept and the idea of tradeable emission permits gave rise to international negotiations, which eventually led to their acceptance as instruments of global environmental policy (see chapter 6). Despite the attention economic instruments received in the 1980s from the scientific, industrial and political world, their use is still limited. Moreover, there is still little evidence about their functioning and effectiveness. From the case studies available (Klok, 1991; Vermeulen, 1992; Ligteringen, 1999) the overall impression is that taxes, levies and subsidies have a rather marginal impact on decision making by target groups, in some cases only supporting environmentally-relevant decisions that have already been taken.

Some authors argue that the instrumentation shift went even further, since not only economic but also social instruments were put into practice to complement the regulatory approach. They rightly refer to the introduction and growing importance of information, communication and the like in environmental policies. These instruments are partly initiated by governmental bodies, underpinning their regulatory policies with information and education campaigns. To a large extent though, information and education campaigns are launched and conducted by non-governmental agencies such as nature conservation organisations, environmental action groups, consumer organisations and others.

1.3.3 The organisation of environmental politics: sharing of responsibility

Other so-called 'communicative' instruments go even further, because they actually imply an organisational change within environmental policies. As stated above, the discussions about the limitations of the regulatory strategies took place within the context of more fundamental political and scientific arguments. These concerned governance in general, the role and responsibilities of the parties involved, and especially the division of responsibility between the state, the market and civil society. Glasbergen (1998) states that, apart from and in reaction to 'regulatory control' as a form of governance, one can distinguish four models of governance, each of which can be regarded as an attempt to control the relations between public and private realms in other ways: market regulation, civil society, contextual control and self-regulation, and co-operative management. The first model assigns a key role to price mechanisms to correct market failures, which have been held responsible for huge environmental problems. The second relies on the social creativity of citizens to overcome environmental deterioration. The third goes even further, by relying upon the self-reflection of societal subsystems. The fourth assigns a key role to the collaborative efforts of governments, private agencies and non-governmental organisations, expecting governance to result from dialogue and communication, in mediation between the parties involved. This last model, incidentally, is expressed in very different arrangements, varying from so-called network management to a variety of public-private partnerships.

Jänicke and Weidner (1995) present a comparable categorisation of governance models when they state that modern industrial societies provide three potential levels of intervention: the traditional government, civil society and the modern 'co-operative state'. Based upon an evaluation of twenty-four case studies from twelve OECD countries, Jänicke and Weidner conclude that these cases reflect a political change which has taken place. This change, described as political modernisation, implies such tasks 'as fiscal reform, management innovation in public administration, better policy co-ordination mechanisms, decentralisation, open input structures in the policy process, institutions for 'anticipatory planning' and long-term strategic orientation. But it also implies political and societal framework conditions that generally create a greater problem-solving capacity and a better infrastructure for the innovator' (Jänicke, 1995: 24).

This comes close to the ecological modernisation theory, which identifies two options for strategies to overcome the deficiencies of the traditional bureaucratic state in environmental policy making. The first option concerns a transformation of state environmental policy making: from curative and reactive to preventive, from exclusive to participatory, from centralised to decentralised, and from over-regulated to stimulating. In short, this option amounts to a modernisation, an updating of the classical repertoire of environmental policy instruments. The second option includes a transfer of responsibilities, tasks and incentives from the state to the market - and, to a lesser extent, to civil society. In this option private agencies, firms and consumers, are all regarded as important social carriers of an ongoing process of socio-ecological transformation (Mol, 1995). According to Spaargaren (1997: 15) the central feature of the ecological modernisation approach as a theory of political modernisation is its focus on new forms of political intervention, especially the changing role of the state. One has to add, however, that the ecological modernisation theory as elaborated by Mol and Spaargaren deals primarily with the state-market interaction, and pays less attention to the changing interrelations of state and market with civil society (cf. Blowers, 1997).

Whether they label it co-operative management, political modernisation or even ecological modernisation, these and other authors all refer to newly-developed forms of governance within contemporary society, and within the environmental field in particular. Behind these new forms of governance lie the changing interrelations between state, market and civil society, which we will discuss in chapter 2. They form the context and are the breeding ground within which the new styles and forms of governance arise and develop. These new forms of governance in turn gradually institutionalise in a wide variety of newly-developing arrangements, one aspect of which is changing instrumentation. In the next chapters we will examine some of these environmental policy arrangements in a more comprehensive way, looking at the actors involved, their resources and power relations, the discourses they use, and so on.

1.4 CONCLUSIONS

In this chapter we have elaborated on the institutionalisation of environmental politics. Institutionalisation, in its sociological meaning, refers to the permanent process of construction and reconstruction of both content and organisation of environmental politics. In section 1.3 we made clear that many studies on the development of environmental politics in Western European countries use, albeit implicitly, an institutionalisation perspective. We paid particular attention to the way they described the changing 'naming' and 'framing' of environmental problems, the shift in the predominant strategies of environmental politics, and in its organisation and division of tasks and responsibilities. These writings provide a comprehensive state of the art as far as the development of Western environmental politics is concerned.

When summarising and drawing some conclusions from these writings, most authors seem to suggest, either implicitly or explicitly, a rather unilinear development in the institutionalisation of environmental politics. That is, first of all, the case as far as the changing social definition of environmental problems is concerned. Almost all authors suggest a gradual development. In the early 1970s environmental problems were defined as local (public health) problems, capable of being solved through technology, with mainly local or at least geographically-restricted impacts. The predominant problem definition in the 1990s, however, can be characterised by stating that problems have diffuse sources and impacts, are difficult if not impossible to solve by technical means, are global in reach and impact, while sustainability, biodiversity and other encompassing concepts are crucial criteria for their assessment. Whereas this might well be the overall direction of development, the unilinear bias might tend to neglect still existing 'hot spots' threatening public health, or underestimate the predominant problem definition of lay people. The problem definition is reflected in and reproduced by the strategies and instruments of early environmental policy. Here again most authors unanimously suggest a rather unilinear development, from simple and *ad hoc* to complex and planned, from sectoral to comprehensive, from fragmented to co-ordinated, from impact-oriented to source-oriented, from regulatory to interactive, from state-initiated to co-operative, and so on. The same unilinear bias applies to their notion of the organisational changes, on which unanimity seems total: from state regulation via market instruments to interactive decision making and co-operative management.

Because most of these writings use an evolutionary perspective and in doing so suggest a too unilinear modernisation of environmental politics, the dynamics and plurality of the process are being underestimated. To overcome this shortcoming we will in the next chapters introduce the idea of political modernisation on the one hand, and the concept of policy arrangements on the other. We believe these concepts offer better opportunities to study the plurality and dynamics of the institutionalisation of environmental politics and policies. Political modernisation refers to major changes in the political field, especially regarding state, market and civil society interrelations. A 'policy arrangement' is understood as the substantive and organisational stabilisation of a policy domain in terms of agents, rules, resources and discourse. They are formed as a result of interactions between actors (and coalitions) in day-to-day policy making, but structured within a specific political context. The combination of the two concepts, therefore, allows us to describe and to understand the specific type of arrangements likely to occur under given political conditions.

REFERENCES

Albrecht, S.L. (1975), The Environment as Social Problem, A.L. Mauss, *Social Problems as social movements*, New York, pp. 556-605.

Bachrach, P. and M. Baratz (1970), *Power and Poverty: theory and practice*, Oxford, New York: Oxford University Press.

Bader, V.M. (1991), *Collectief handelen: sociale ongelijkheid en collectief handelen*, Groningen: Wolters-Noordhoff.

Baumol, W.J. and W.E. Oates (1975), *The theory of environmental policy*, Englewood Cliffs, N.J.: Prentice-Hall.

Blowers, A. and P. Leroy (1996), Environment and Society: shaping the future, Blowers, A. and P. Glasbergen eds., *Environmental Policy in an International Context, book 3 - Prospects*, London, pp. 255-283

Blowers, A. (1997), Environmental Policy - Ecological Modernization or the Risk Society, *Urban Studies*, 34(5/6), 845-871.

Buttel, F.H., A.P. Hawkins, et al. (1990), From Limits to Growth to Global Change - Constraints and Contradictions in the Evolution of Environmental Science and Ideology, *Global Environmental Change*, 1(1), 57-66.

Buttel, F.H. and P. Taylor (1992), Environmental sociology and global environmental change - A critical assessment, *Society and Natural Resources*, (5), pp. 211-230.

Cobb, R.W. and C.D. Elder (1972), *Participation in American Politics: The Dynamics of Agenda-Building*, Boston: Allyn and Bacon.

Cohen, M.D., J.G. March, and J.P. Olson (1972), A Garbage Can Model of Organizational Choice, *Administrative Science Quarterly*, 17(1), pp. 1-25.

Downs, A. (1972), Up and Down with Ecology. The 'issue-attention cycle', *Public Interest*, 28, pp. 38-50.

Eijk, C. van der and W.J.P. Kok (1975), Non-decisions reconsidered, *Acta Politica*, 10, pp. 277-301.

Fisher, R. and J. Kling (1993), *Mobilizing the community*, Newbury park: Sage.

Garner, R. (1996), *Environmental Politics*, London: Prentice Hall, Harvester Wheatsheaf.

Glasbergen, P. eds. (1998), *Co-operative Environmental Governance. Public-Private Agreements as a Policy Strategy*, Dordrecht: Kluwer Academic Publishers

Hajer, M. (1997), *The Politics of Environmental Discourse. Ecological Modernization and the Policy Process*, Oxford, New York: Oxford University Press, Clarendon Press.

Hanf, K. (1996), Implementing International Environmental Policies, A. Blowers and P. Glasbergen (eds.), *Environmental Policy in and International Context III. Prospects*, London: Arnold, pp. 197-222.

Hannigan, J.A. (1995), *Environmental Sociology. A social constructionist perspective*, London and New York: Routledge.

Jänicke, M. and H. Weidner, eds. (1995), *Successful Environmental Policy. A Critical Evaluation of 24 Cases*, Berlin: Edition Sigma.

Kingdon, J.W. (1984), *Agendas, Alternatives and Public Policies*, Boston, Toronto: Little, Brown and Company.

Klok, P.J. (1991), *Een instrumententheorie voor milieubeleid: de toepassing en effectiviteit van beleidsinstrumenten*, Enschede: Universiteit Twente, Faculteit Bestuurskunde.

Liberatore, A. (1995), The Social Construction of Environmental Problems, P. Glasbergen and A. Blowers, *Environmental Policy in an International Context, Tome 1 - Perspectives*, London, New York: Wiley, pp. 59-84.

Ligteringen, J.L. (1999), *The feasibility of Dutch environmental policy instruments*, Enschede.

March, J.G. and J.P. Olson (1984), The new institutionalism: organizational factors in political life, *The American Political Science Review*, 78(3), pp. 734-749.

Mayntz, R. (1978), *Vollzugsprobleme der Umweltpolitik. Empirische Untersuchung der Implementation von Gesetzen im Bereich der Lufteinhaltung und des Gewaesserschutzes*, Stuttgart: Kohlhammer.

Meadows, D.L. (1972), *Rapport aan de Club van Rome. De grenzen aan de groei*, Utrecht, Antwerpen: Spectrum.

Mol, A.P.J. (1995), *The Refinement of Production. Ecological modernization theory and the chemical industry*, Utrecht: Van Arkel.

Opschoor, J.B. and H.B. Vos (1989), Economic instruments for environmental protection, Paris: OECD.

Potter, D. (1995), Environmental problems in their political context, Blowers, A. and P. Glasbergen eds., *Environmental Policy in an International Context, book 1 - Perspectives*, London, pp. 85-110

Pressman, J.L. and A. Wildavsky (1973), *Implementation: how great expectations in Washington are dashed in Oakland; or, Why it's amazing that Federal programs work at all, this being a saga of the Economic Development Administration as told by two sympathetic observers who seek to build morals on a foundation of ruined hopes*, Berkeley: University of California Press.

Schoenfeld, A.C., R.F. Meier and R.J. Griffin (1979), Constructing a social problem: the press and the environment, *Social Problems*, 27(1), pp. 38-61.

Spaargaren, G. (1997), *The Ecological Modernization of Production and Consumption. Essays in Environmental Sociology*, Wageningen: WAU.

Steunenberg, B., J. de Vries, and J.L. Soeters (1996), Het neo-institutionalisme in de bestuurskunde, *Bestuurskunde*, 5(5), pp. 212-216.

Van Tatenhove, J. (1993), *Milieubeleid onder dak? Beleidsvoeringsprocessen in het Nederlandse milieubeleid in de periode 1979-1990; nader uitgewerkt voor de Gelderse Vallei*, Wageningen: Pudoc.

Vermeulen, W. (1992), *De vervuiler betaald. Onderzoek naar de werking van subsidies op vier deelterreinen van het milieubeleid*, Utrecht: Van Arkel.

Weale, A. (1992), *The new politics of pollution*, Manchester and New York: Manchester University Press.

Wells, D.T. (1996), *Environmental Policy: A Global Perspective for the Twenty-First Century*, New Jersey: Prentice Hall.

Yearley, S. (1992), *The Green Case. A sociology of environmental issues, arguments and politics*, London, New York: Routledge.

Zald, M.N. and J.D. McCarthy, eds. (1979), *The dynamics of social movements: resource mobilization, social control, and tactics*, Cambridge, Mass.: Winthrop.

2. Political Modernisation

JAN VAN TATENHOVE, BAS ARTS AND PIETER LEROY

2.1 INTRODUCTION

In this and in the next chapter we introduce the concepts of political modernisation and policy arrangements, in order to study the plurality and dynamics of processes of institutionalisation in environmental politics. Institutionalisation not only relates to how day-to-day policy making is structured by policy arrangements which result in policy, but also to how the plurality of policy arrangements develops within a certain political context and in a specific period of time. The kind of arrangements and the plurality of these arrangements, we will argue, result from long-term processes of transformation within the political domain, which we will describe as political modernisation.

In section 2.2 we distinguish three stages of political modernisation: 'early', 'anti-' and 'late'. Each of these stages can be characterised by specific interrelations between state, civil society and market, and by specific types of governance. Empirically the changing interrelations between state, civil society and market can be observed in the tendency in many countries to shift the balance between government and society away from the public sector and more towards the private sector. These new patterns of interaction, but also new divisions of labour and balances of power, are apparently aimed at discovering fresh ways of coping with new problems or at creating new possibilities of governing (Kooiman, 1993a). In general, governance is a term that applies to the exercise of power in a variety of institutional contexts, the object of which is to direct, control and regulate activities in the interest of people as citizens, voters and workers (Kuper and Kuper, 1996). In discussing new patterns of interaction between government and society Kooiman discerns two levels of analysis. On the concrete governing level there is the search for new models of governing in terms of jointly, such as co-steering, co-managing, co-producing and coallocating. 'But one can also observe changes in terms of patterns of governance: broader and maybe more pervasive efforts to come to grips with fundamental developments and structural characteristics of the societies we live in' (Kooiman, 1993a: 2). In this book we also use a dynamic concept of governance, 'balancing' between different levels of governing and governance (compare Kooiman, 1993b: 43-47) which result from a particular division of power between predominant actors, through specific discourses on governance, and through changing interrelations between state, civil society and market.

The central line of argument on political modernisation is that each phase can be typified by specific discourses on governance and characteristic interrelations between state, civil society and market. As a result of these discourses and relationships between state, civil society and market, new actors emerge while other actors' roles change, interdependencies among societal and political actors change, new discourses gain more legitimacy and power, coalition and domination relationships are modified. Based on this argument, we will in chapter 3 discuss the types of coalitions, the legitimate discourse and the rules and resources which form specific policy arrangements on the crossroads of processes of social and political change and day-tot-day policy making.

2.2 POLITICAL MODERNISATION

In general modernity refers to modes of social life or organisation which emerged in Western countries from about the seventeenth century onwards, the influence of which subsequently became more or less worldwide. It is a cluster of cultural and structural processes, of typical institutional forms, which came into being on the crossroads of capitalism, industrialism and the nation state (cf. Giddens, 1990). Political modernisation, then, refers to processes of transformation within the political domain of society. It is not easy to demarcate the political domain of society exactly. Does it concern the interplay between government, parliament and administration only? Are interest groups and citizens also part of it? In general, the demarcation of the political domain of society depends on the degree of insulation of the subsystems state, market and civil society. In a situation where a sharp distinction between state, civil society and market exists, the position of the political domain will be mainly defined in terms of the rationalities of the state. However, contemporary societies show increasing encroachment, interweaving and interference of the three subsystems and demarcation lines are rather vague. In this book we therefore deliberately use a broad concept of the 'political domain' (cf. Held, 1989). Essentially, the political domain of society is the setting in which different groups (from state, civil society and market) produce and distribute resources (power and domination) and meaning (discourses) to shape public life. Structures of domination always have to be understood in connection with structures of signification and legitimation. The former refers to the mobilisation of both allocative and authoritative resources, whereas the latter refers to modes of meaning constitution, and normative sanctions (cf. Giddens, 1984). For example, how do the rationalities of bureaucracy, competition and solidarity come together in the political domain of society? And how are structures of domination, legitimation and signification produced and reproduced in concrete policy making?

To grasp the dynamics of the process of political modernisation we distinguish analytically between three phases of political modernisation: 'early', 'anti-' and 'late' (cf. Alexander, 1995). In the context of this chapter it is neither possible nor necessary to discuss all theories of post- and reflexive modernity and modernisation in detail. We therefore use elements of the debate and summarise these under the label of 'late political modernisation'. For extensive overviews we refer to others (Lash and Urry, 1987; Mommaas, 1993; Wagner, 1994; Kumar, 1995). Each of the three phases mentioned can be characterised by specific, ideal-typical relations between state, civil society and market, and by dominant discourses on governance. In this section we concentrate on the changing discourses on governance, and on the changing interrelations between state, market and civil society in each of the phases. In our view, these long-term changes affect the characteristics of the types of policy arrangements prevailing in each period. The concept of policy arrangements will be discussed in chapter 3.

2.2.1 Early Political Modernisation

Discourses on governance

The 'early political modernisation' phase is closely linked to the project of modernity itself. For an extensive analysis of the characteristics of the 'early' or 'simple' modernisation model, see among others Giddens (1990), Alexander (1995), Kumar (1995) and Albrow (1996). As we cannot, within this chapter, give a detailed analysis on discourses of modernity, we will restrict ourselves to some aspects only. Central elements of the early (political) modernisation discourse are the progress of control over both society and nature. This is not only a matter of ideas, as Giddens (1990) makes clear when defining modernity as a cluster of institutional forms and processes, which come into being at the crossroads of capitalism, industrialism and the nation state. Although modernity itself is a permanent revolution of ideas and institutions (see Kumar, 1995; Wagner, 1994), these elements come together in the notion of the 'manageable society'. The 'manageable society' refers to the idea of man's ability to shape both the social and the physical world. The relevance of this concept to environmental issues is self-evident. The control and exploitation of the physical environment are demonstrated in the increasing depletion of natural resources and in the dumping of waste into the natural environment, resulting in a deterioration of the 'sustenance base' of society (Schnaiberg, 1980).

Albrow typifies the project of modernisation as linking the world by a twofold process. First, the imposition of practical rationality upon the rest of the world through the agency of the state and the mechanisms of the market, and secondly the generation of universal ideas to encompass the diversity of the world (Albrow, 1996: 33). This project is characterised by rationalisation and control of human interactions and interdependencies, mainly through the institutions and policies of the (welfare) state. According to Albrow (1996: 189),

> '(...) underpinning modern discourse was the assumption that there were rational ways to determine optimum arrangements of the elements and that these necessarily integrated the spheres of individual and state activity and all the intermediate instances of social activity, such as family, firm and community. The principle of hierarchy appeared in numerous ways, in decision-making trees, in relations of authority, in levels of abstraction or of jurisdiction'.

Although the modernisation project was driven by a number of different ideologies, they shared the basic assumption that it is possible and recommendable to outline the most desirable development of society. In the political domain this discourse of the constructed society is translated and reflected in models of rational policy making and synoptic planning, pluralistic concepts of democracy and functional concepts of steering and governability. Rational choice theories are predominant and rational policy makers play a central role within the political domain. Public actors interacting within the domain of the state design - in a rational manner - a complex of coherent means and ends. Policy makers try to solve problems along the lines of scientific rationality, and processes of policy making are organised around rational scientific principles. One of these principles is specialisation, leading among other things to an increasing differentiation in both state legislation and public administration, including the emergence and proliferation of governmental bodies when new problems arise. As we have seen in the Introduction and chapter 1 these processes of specialisation, differentiation and fragmentation affected environmental policies in specific ways.

Relations between state, market and civil society

The predominant concept in the 'early political modernisation' stage was that of a relative insulation of state, market and civil society, with each sphere functioning according to its own rationales: bureaucracy, competition and solidarity respectively. Nevertheless, the state was regarded as the allocative and authoritative 'power container'. As a consequence the influence of private actors, either market or civil society agents, on the process of policy making was limited in practice and neglected or underestimated in theory. This notion agreed with ideas on governance and governability which were dominated by assumptions about a monocentric state, steering civil society more or less as if it were a subject-object relationship, and expected to be able to construct 'the good society'.

However, this concept of a relative insulation of the subsystems state, market and civil society, with their own distinctive rationales, does not imply a fixed position of each one in relation to the others. On the contrary, specific interrelations between state, market and civil society depend upon the predominant political preferences of a society at a given time, and upon its political and ideological traditions and their institutional structuring, which give rise to different policy arrangements (see chapter 3). Examples of such arrangements, also implying specific concepts of governance, governability and policy making, are statism, corporatism and liberalism (Frouws, 1993; Leroy and Van Tatenhove, 2000; Williamson, 1989). In statist arrangements the political institutions are the ultimate locus of authoritative power and they therefore largely determine the contents and the organisation of policies. In corporatist arrangements political authority is shared by the state and some acknowledged intermediate organisations of 'stakeholders' in certain policy domains, especially trade unions and employers' organisations (e.g. social security and welfare policies). Liberal arrangements are characterised by the domination of market agents. From this the empirical question arises as to which institutional arrangement models have predominantly inspired environmental policy when it emerged in the early 1970s, during its development in the 1980s, and during its renewal in the 1990s. This is the central focus of this book, for which we construct an analytical framework in this and the next chapter. In chapters 4 to 8 we then present empirical evidence on the institutionalisation of subdomains of environmental policy.

Due to processes of regionalisation and (early) globalisation, the subsystems state, market and civil society have been internationalised since Second World War, at least to some extent - there is, for example, no such thing as 'the global state'. Although policy arrangements at international level are of a rather different nature to domestic ones, and the above notions - statist, corporatist, liberal - are therefore less applicable, different arrangements can be distinguished at the European and the global levels as well. Traditionally, the 'intergovernmental' arrangement is predominant: in these arrangements state agents - heads of state, prime ministers, diplomats, government representatives - determine political outcomes. This dominance of nation states is 'the premise' of international law (Hocking and Smith, 1990; Ray, 1987). However, private organisations have played (minor) roles in international policy making for a long time, and did so even in the days of the League of Nations (Woods, 1993). Since the Second World War, though, the position of non-state actors - NGOs, business lobbies, 'epistemic communities' etc. - has certainly been strengthened, both in the EU and the UN. Simultaneously the power of these international organisations has also increased. Authors therefore refer to both a diachronic and a synchronic development from 'intergovernmental' to 'transnational' arrangements in the international arena in which state, market and civil agents co-determine political outcomes (Hogenboom, 1998; Keohane and Nye, 1971; Keohane and Nye, 1989; Reinalda, 1997; Risse-Kappen, 1995).

2.2.2 Political Anti-Modernisation

Discourses on Governance

According to Alexander (1995), at some point in the late 1960s modernisation theory died. Although others doubt this is really the case (cf. Giddens, 1990; Bauman, 1992; Beck, 1994, Beck, 1996; Kumar, 1995; etc.), Alexander's notion of an anti-modernisation discourse offers an interesting clue in explaining one of the trends within the project of political modernisation. The 1970s show both a renewal of theories of modernisation and the emergence of radical alternatives, the latter being theories and ideas of anti-modernisation. Alexander rightly sees anti-modernisation as a reaction to the unsolved 'reality problems' of the modernisation project, such as inequity and inequality, poverty and starvation, dictatorship and postcolonialism. In the 1970s theories of anti-modernisation were proposed as providing more valid explanations of these problems. They focused on issues such as inequality and emancipation, democracy and participation. Both the newly-developing social movements of the early 1970s and the academics and intellectuals which inspired them, characterised the period of 'early political modernisation' as one-dimensional, materialistic, bureaucratic and repressive (Marcuse, 1964; Habermas, 1968). In their opinion modern Western society was not rational and liberating, but backward, greedy, oppressive and impoverishing.

Central elements of the anti-modernisation discourse, such as politicisation and emancipation, were elaborated on in neo-Marxist, eco-socialist, counter-productivity, dependence and de-industrialisation theories (cf. Frank, 1969; Bahro, 1982; Huber, 1985; Keohane, 1989; Sachs, 1977; Ullrich, 1979). These discourses were added to by the new social movements of the late 1960s and early 1970s, focusing on Third World issues, the environment and other political issues (see Jamison, Eyerman and Cramer, 1990). Both critical social theorists and new social movements played a crucial, explicit role in the formulation of discourses opposed to capitalism and to the oppressing role of the state. It is striking to see all these ideals - the emancipation of the underprivileged, the liberation of labour from capitalist alienation etc. - being defined in terms of conflict, revolution and collective emancipation, combining the ideas of subversion of the state and politicisation of civil society (cf. Offe, 1986). The aim was to establish totally new linkages between the state and civil society, corresponding to classical ideas of 'the political community'. Therefore, while the state was under attack and anti-modernisation theorists emphasised the legitimation crisis of the state, some theories were still very state-centred, aiming for a takeover of state institutions by societal organisations (e.g. Bahro, 1982). Other authors, though, returned to the ideas of pre-Marxist or utopian socialism, planning the establishment of small-scale communities which are politically sovereign and economically autarkic, with small-scale technologies, thus enabling a 'communal society' (e.g. Illich). Some authors depicted this political revolution as accompanied by a (partial) dismantling of the existing economic and technological systems of production (e.g. Commoner, Roszak, Gorz, Ullrich, Sachs). Counter-productivity and de-industrialisation were part of a project of de-modernising society. A few authors went even further, in outlining an anarchistic renewal of society and politics (e.g. Bookchin).

It is not surprising that environmental issues were at the centre of many of these theories and thoughts, as they seemed to be the example *par excellence* of the one-dimensional early modernisation. Two main characteristics of the modernisation project, the strict separation of facts and values on the one hand and the unceasing process of specialisation on the other, resulted in rationalistic reductionism, which offered science, technology and state administration alike a very limited view of reality. Classical economics, for example, and the policies based upon it devoted no attention to the adverse consequences of economic growth, as Mishan pointed out in *The costs of economic growth*. Specialisation, therefore, narrowed the scope in disregarding the so-called externalities, one of the symbolic catchwords of the anti-modernisation thoughts. Environmental issues were one example of neglected side effects, neo-colonialism, the arms race and the atom bomb, to name but a few, were others. These anti-modernisation theories to a large extent furthered environmental concern among the Western public, and as such played a crucial role in the emergence of environmental policies. Although anti-modernisation theorists firmly opposed the thesis of early modernisation, they still believed in the grand narrative of the 'manageable society'. The proposals of the environmental movement in particular, forced governments to reflect continuously on the substance and organisation of environmental politics.

Relations between state, market and civil society

It is clear from the above that the anti-modernist criticism focused on the one-sided, one-dimensional and reductionist role of the state and the market, perceived as a *de facto* coalition of power, which reinforced the social order in favour of particular capitalist interests, both at national and international level. Although the criticism and the proposals differed in many respects, as we have indicated, there were some common points, of which the key one was a strong opposition to the domination of the coalition of the capitalist state and the market over civil society. As an alternative pleas for social and political emancipation of the underprivileged were launched, which would result in forms of radical democratisation and self-government, either on the local, the national or on the global level. In short, a reversal of the existing social and political order was called for. However, none of these proposals for a different social and political order has been implemented in the Western world. Nonetheless these visions influenced politics and policy making on the ground. By launching new concepts they, firstly, provoked a change in the content of politics, resulting in the emergence of well-considered policies on postmaterial issues such as the environment, education, culture, community work and other newly-developing forms of welfare policies. Secondly, within these newly-developing policy domains, their programs and their projects, new policy arrangements were created, characterised by participation, self-regulation and self-determination. And thirdly, similar experiments and changes took place within some traditional policy domains as well, through the acknowledgement of new political actors, the recognition of new problem definitions and policy discourses, and so on, leading to the (partial) renewal of traditional policy arrangements.

2.2.3 Late Political Modernisation: an Epochal Shift?

Discourses on Governance

Contemporary society is in transition, but the direction of societal and political changes is the subject of heated debates among social scientists. Under the banner of post-traditional, post-Fordism, post-industrial, post-, reflexive and late modernity and so forth, contemporary changes are analysed and understood in many different ways. In the context of this chapter we cannot give an overview of the different positions in the debate, but we will discuss post- and reflexive modernisation in general, using elements of these debates to typify what we will call the phase of late political modernisation. The prefix 'late' refers to the general idea that the 'manageable society' is being redefined, mainly as a consequence of two basic processes of change: globalisation and individualisation. Globalisation and individualisation are two major changes within the political, economic and socio-cultural domains. Although their empirical extent and their actual impact have been challenged (cf. Hirst and Thompson, 1996), these processes are said to provoke a new stage in modernisation in general and in political modernisation in particular. In political modernisation these changes are reflected by significant changes in governance and changing relations between state, civil society and market.

In general, globalisation implies that many political, economic and social activities are becoming worldwide in scope. At the same time it suggests that there has been an intensification of levels of interaction and interconnectedness within and between states and societies (Held, 1995: 21). Policy making and governance will be greatly influenced, no matter whether homogeneous or heterogeneous tendencies of globalisation are accentuated (Robertson, 1992). The first refers to developments in standardisation, unification and universalisation of social, economic, political and cultural patterns on a global scale, the latter emphasises the diversity of reactions to these homogeneous tendencies.

Individualisation is the second major change, making late, post- or reflexive modernity the era of the individual, linked to the (globalised) world through a mobile phone and the Internet. Many authors describe individualisation as an all-pervasive process, made possible by modern technologies and opening new perspectives for emancipation. At the same time, though, individualisation leaves the individual unprotected either by the pre-modern 'mechanic solidarity' of family, kinship, neighbourhood, local community, church etc., or by the modern 'organic solidarity' of welfare institutions of the provident state. The former were eroded during the early modernisation stages, whereas the latter have been deregulated or privatised during the wave of 'less state' politics that has dominated Western countries over the last two decades or so - a wave that has been largely 'globalised', incidentally. The erosion of kinship solidarity and the withdrawal of the provident state are threatening the individual with personal insecurity (in cases of disease, relationship problems and the like), and economic insecurity (in the case of unemployment, changing conditions of employment etc.).

Because post- and reflexive modernity presuppose a different relationship to the project of modernity, they interpret globalisation and individualisation in different ways. In general, post-modernists believe we are facing a radical transformation of modernity. The Modern project has ended and, according to Albrow, has been transformed into the *Global Age*, which involves the supplanting of modernity by globality: an overall change in the basis of action and social organisation for individuals and groups. 'Globality is a new level of organisation, to which any agent can relate, but which has no organising agent' (Albrow, 1996: 121). According to some post-modernist theorists contemporary societies show a new or intensified degree of fragmentation, pluralism and individualism, while political, economic and cultural life is strongly influenced by developments at the global level. Post-modernism proclaims multicultural and multiethnic societies and promotes the 'politics of difference' - linking the local and the global - in which identity is not unitary or essential but fluid and shifting, fed by multiple sources and taking multiple forms (Kumar, 1995). The post-modern world is summarised by (Kumar, 1995: 147-148) as

> 'a world of eternal presentness, without origin or destination, past or future; a world in which it is impossible to find a centre, or any point or perspective from which it is possible to view it steadily and to view it whole; a world in which all that presents itself are the temporary, shifting and local forms of knowledge and experience. Here are no deep structures, no secret or final causes; all is (or is not) what it appears on the surface. It is an end to modernity, and all that is promised and proposed'.

Other scholars, advocates of the reflexive modernisation thesis, however, understand contemporary changes as an accumulation of and within modernity (e.g. Giddens, Beck, and Castells). In reflexive modernisation the globalisation of risks is put at the forefront. Moreover, the individual is threatened by new risks, like BSE, the E-coli bacillus, and global risks, such as global warming. It seems as though there is no way to overcome the global risks, but at the same time individuals have to live with these unforeseen consequences of modernity. 'Individualisation (...) means that each person's biography is removed from given determinations and placed in his or her hands, open and dependent on decisions' (Beck, 1992: 135). Beck therefore depicts contemporary society as a *risk society*, which he assumes to be a stage of 'high' or 'radicalised' modernity. Essential here is the unintentional and unseen transition from modern industrial society to risk society. This risk society seems to be dominated by uncontrollable economic-technological developments. That is especially true for the so-called high-consequence risks, defined as risks, beyond the control of individual agents while at the same time threatening the lives of thousands if not millions, arising from chemical engineering, nuclear technology, biotechnology etc. The risk society is understood to be a phase of development of modern society in which the amalgam of individual, social, economic and ecological risks, created by the momentum of innovation, increasingly eludes the control and protective institutions of industrial society. Like the individual, society as a whole also can no longer rely upon traditional institutions, and 'the less we can rely on traditional securities, the more risks we have to negotiate. The more risks, the more decisions and choices we have to make' (Beck, 1998: 10). At the same time, though, Beck depicts the risk society as a well of pessimism, fatalism and despair, when he states that the risk society is not an option to be chosen or rejected in the course of political debate (Beck, 1996: 27-28).

Although Beck maintains that the risk society has 'altered the epistemological and cultural status of science and the constitution of politics' (Beck, 1998: 10), he does not elaborate on the controllability of the risk society. According to his central argument, the side effects of modernisation, especially the globalisation of environmental risks, have become the pivot of governance. Apart from 'subpolitics', which we deal with below, 'reflexive modernisation' seems to be the only solution. Reflexive modernisation for Beck means 'self-confrontation with the effects of risk society that cannot be dealt with and assimilated in the system of industrial society - as measured by the latter's institutionalised standards' (Beck, 1994: 6). Therefore, new ways of governance have to be developed within and beyond the nation state model (Wapner, 1995).

Relations between state, market and civil society

Whatever post- and reflexive modernity may imply, it is clear from the above that globalisation, individualisation and the emergence of the risk society affect the nation state model. In fact, this model seems to have lost its exclusiveness, as it is paralleled by a whole series of local, regional and global arrangements, set up by actors from different spheres and domains, crossing the traditional borders of nation states and their divides. Examples are transboundary co-operation between districts, competition between cities and regions, grouping of nation states, and the emergence of global cities as new economic power structures. At the same time the relations between state, market and civil society become less clear, because they have been affected by both heterogeneous and homogeneous processes of globalisation, and by the interrelations of actors crossing the boundaries of these spheres, which were previously more or less insulated. Individualisation, as we have indicated, deeply affects the institutions within civil society and those linking it to both the state and the market.

In Beck's view, the controllability of the whole is problematic, as the risk society becomes a laboratory in which nobody is responsible for the outcomes of experiments (Beck, 1998: 10):

> 'The private sphere's creation of risks means that it can no longer be considered apolitical. Indeed, a whole arena of hybrid sub-politics emerges in the realms of investment decisions, product development, plant management and scientific research priorities. In this situation, the conventional political forces and representations of industrial society have been sidelined'.

According to Beck, the essence of contemporary politics is (reflexive) sub-politicisation of society.

> Sub-politics 'means shaping society *from below*. Viewed from above, this results in the loss of implementation power, the shrinkage and minimization of politics. In the wake of sub-politicisation, there are growing opportunities to have a voice and a share in the arrangement of society for groups hitherto uninvolved in the substantive technification and industrialisation process: citizens, the public sphere, social movements, expert groups, working people on site; there are even opportunities for courageous individuals to 'move mountains' in the nerve centres of development. Politicization thus implies a decrease of the central rule approach; it means that processes which had heretofore always run friction-free fizzle out in the resistance of contradictory objectives' (Beck, 1994: 23).

In his interpretation of sub-politics Beck makes a distinction between 'rule-directed' and 'rule-altering politics'. Rule-directed politics functions within the rule system of the nation state, in itself a result of what Beck calls the 'first' modernity. Rule-altering politics on the other hand concerns changing the rules of the game. In reflexive political modernisation rule-directed and rule-altering politics mingle, overlapping and interfering with each other. According to Beck, an essential element of 'directed politics' is the formation of *ad hoc* 'coalitions of opposites', in opposition to modern institutions. In his elaboration of globality Albrow also highlights these new coalitions. Since globality for him is associated with a recovery of *free sociality*, with the expression of a global public opinion and the mobilisation of global movements, he introduces the concept of *the global state*. The global state exists at 'every moment when the individual takes account of and seeks to perform in the interests of a common interest spanning the globe' (Albrow, 1996: 178). Like Beck's sub-politics, the global state is also constructed from below upwards and relies on the global consciousness of countless individuals. In this context Albrow emphasises 'performative citizenship': the expression of concern for public good in new configurations (depending on constructed identities beyond the nation state), which are gradually beginning to challenge nation states for the loyalties of their citizens.

While it may be true that in this phase of late political modernity the nation state model has lost its exclusiveness, and that the traditional divides between the state, the market and civil society have been crossed, this does not imply an unambiguous development. Not only do post- and reflexive modernist advocates judge contemporary transitions in societies in different ways, there are also conservative analyses which come to similar points of view, although from an almost directly opposite angle. The neo-conservative discourse, for example, also stresses the formation of a smaller but stronger state; emphasises the problem-solving capabilities of other agencies, especially private market agents; and relies to a great extent upon the self-steering abilities of civil society. The first led to the withdrawal of the state from many domains, the second to various forms of privatisation of former state tasks, to their devolution to quangos (quasi autonomous non-governmental organisations), or to several kinds of public-private partnership, while the third led to the development of new regimes of governance, characterised as negotiated decision making, communicative governance, co-operative management and so forth.

Essential to the phase of late political modernity is its plurality. The formation of new coalitions, the allocation of resources, the way actors construct identity, is perceived both within and outside the nation state model, and influenced by local-global interconnectedness. These processes will be reflected in the formation and construction of policy arrangements. We will discuss this plurality of new policy arrangements in general in the next chapter and expand on it when we present the empirical data from different domains of environmental policy in chapters 3 to 8. Central questions to answer will concern the extent to which the emergence of these new arrangements can be understood in terms of late political modernisation, or whether they have to be seen as the result of a modernist or anti-modernist transformation of the political domain. Whether one regards these arrangements as a sign of post- or reflexive political modernisation or as the vanguard of a neo-conservative reconquest, they imply a shift from a rather limited set of arrangements (statist, corporatist or liberal) towards a wide variety of arrangements of very different hues. The interesting point then is to discover what these new arrangements actually involve: in terms of political participation, in terms of the allocation of political power, in terms of their legitimacy, in terms of their controllability, and in terms of their effectiveness.

2.3 CONCLUSIONS

Processes of globalisation and individualisation undoubtedly affect and will continue to affect the political domain. Although the distinction between the early, the anti- and the late modernist stages of political modernisation is analytical only, so that stages overlap in time, it enables the identification of contemporary changes in the political domain. Essential for this is that (a) the traditional divides between state, market and civil society are disappearing, while (b) the interrelations between these spheres increasingly exceed the nation state, (c) resulting in new coalitions between state agencies, market agents and civic parties both on local and global levels. This last is what Wapner (1995) calls 'world civic politics beyond the state' and what Castells terms the network society. Castells' analysis of 'the network society' is a very good illustration of the new interrelations between state, civil society and market and the new coalitions which result from the tension between globalisation and identity. Or, as he calls it, between the Net (the rise of the global informational economy shaped by the restructuring of the capitalist mode of production) and the Self (the pre-eminence of identity as the organising principle). Networks are analysed as interconnected nodes, such as for example stock markets, national councils of ministers and European Commissioners, coca fields and poppy fields, or television systems, which are able to expand without limits, integrating new nodes as long as they share the same communication codes. The analysis makes it possible to focus on a new dualism of the capitalist society: the 'space of flows' and the 'space of places'. In the network society 'dominant functions are organised in networks pertaining to a space of flows that links them up around the world, while fragmenting subordinate functions, and people, in the multiple space of places, made of locales increasingly segregated and disconnected from each other' (Castells, 1996: 476).

Recent debates make clear that, at the end of this millennium, the relations between state, market and civil society have changed fundamentally. Such change can only be understood in the context of global interconnectedness. This not only refers to the worldwide scope of political, economic and social activities, but also to new levels of interaction, new forms of identity and new sets of meaning. As a consequence, power is no longer concentrated only in modern institutions (state, capitalist firms, corporate media etc.), but is diffused in global networks of wealth, power, information and images, which circulate and transmute in a system of variable geometry and dematerialised geography. New power lies 'in the codes of information and in the images of representation around which societies organize their institutions, and people build their lives, and decide their behavior. The sites of this power are people's minds' (Castells, 1997: 359).

In short, whether regarded as late, post- or reflexive modernity, whether perceived as caused by globalisation, individualisation or other processes, there seems a broad consensus on some recent social changes, which we conceive of as distinguishable stages of political modernisation. It is within this context that we now turn to the renewal of policy arrangements.

REFERENCES

Albrow, M. (1996), *The Global Age. State and Society Beyond Modernity*, Oxford: Polity Press.

Alexander, J.C. (1995), *Fin de Siecle Social Theory. Relativism, Reduction, and the Problem of Reason*, London, New York: Verso.

Bahro, R. (1982), *Elementen voor een nieuwe politiek*, Amsterdam: Van Gennip.

Bauman, Z. (1992), *Intimations of Postmodernity*, London and New York: Routledge.

Beck, U. (1992), *Risk Society: towards a New Modernity*, London: SAGE.

Beck, U. (1994), The Reinvention of Politics: Towards a Theory of Reflexive Modernization, U. Beck, A. Giddens and S. Lash eds., *Reflexive Modernization. Politics, tradition and aesthetics in the modern social order*, Oxford: Polity Press, pp. 1-55.

Beck, U. (1996), World Risk Society as Cosmopolitan Society? Ecological Questions in a Framework of Manufactured Uncertainties, *Theory, Culture & Society*, 13(4), pp. 1-32.

Beck, U. (1998), Politics of Risk Society, J. Franklin, eds., *The Politics of Risk Society*, Cambridge: Polity Press, pp. 9-22.

Bookchin, M. (1980), *Toward an Ecological Society*, Quebec: Black Rose Books.

Castells, M. (1996), *The Information Age: Economy, Society and Culture, Volume I, The Rise of the Network Society*, Oxford UK, Malden USA: Blackwell.

Castells, M. (1997), *The Information Age: Economy, Society and Culture, Volume II, Power of Identity*, Oxford UK, Malden USA: Blackwell.

Frank, A.G. (1969), *Sociology of development and underdevelopment of sociology*, Stockholm: FARSTA-Offset.

Frouws, J. (1993), *Mest en macht. Een politiek-sociologische studie naar belangenbehartiging en beleidsvorming inzake de mestproblematiek in Nederland vanaf 1970*, Wageningen.

Giddens, A. (1984), *The Constitution of Society. Outline of the Theory of Structuration*, Cambridge: Polity Press.

Giddens, A. (1990), *The Consequences of Modernity*, Oxford: Polity Press.

Habermas, J. (1968), *Techniek und Wissenschaft als Ideologie*, Frankfurt: Surkamp Verlag.

Held, D. (1989), *Political Theory and the Modern State. Essays on State, Power and Democracy*, Cambridge: Polity Press.

Held, D. (1995), *Democracy and the Global Order. From the Modern State to Cosmopolitan Governance*, Cambridge: Polity Press.

Hirst, P. and G. Thompson (1996), *Globalization in Question. The International Economy and the Possibilities of Governance*, Cambridge, Oxford: Polity Press.

Hocking, B. and M. Smith (1990), *World Politics. An introduction to international relations*, New York: Harvester/Wheatsheaf.

Hoogenboom, B. (1998), *Mexico and the NAFTA Environment Debate. The Transnational Politics of Economic Integration*, Utrecht: Van Arkel/International Books.

Huber, J. (1985), Ecologische modernisering, E. Van den Abbeele, eds., *Ontmanteling van de groei. Leesboek over een andere economie*, Nijmegen: Markant, pp. 161-168.

Jamison, A., R. Eyerman and J. Cramer (1990), *The making of the new environmental consciousness: a comparative study of the environmental movements in Sweden, Denmark and the Netherlands*, Edinburgh: Edinburgh University Press.

Keohane, R.O. and J.S. Nye, eds. (1971), *Transnational Relations and World Politics*, Cambridge: Harvard University Press.

Keohane, R.O. and J.S. Nye, eds. (1989), *Power and Interdependence. World Politics in Transition*, Glenview: Scott, Foresman & Company.

Kooiman, J. (1993a), Social-Political Governance: Introduction, J. Kooiman, eds., *Modern Governance. New Government-Society Interactions*, London: Sage, pp. 1-6.

Kooiman, J. (1993b), Governance and Governability: Using Complexity, Dynamics and Diversity, J. Kooiman, eds., *Modern Governance. New Government-Society Interactions*, London: Sage, pp. 35-48.

Kumar, K. (1995), *From Post-industrial to post-modern society. New Theories of the Contemporary World*, Oxford UK, Cambridge USA: Blackwell.

Kuper, A. and J. Kuper, eds. (1996), *The Social Science Encyclopedia*, London and New York: Routledge.
Lash, S. and J. Urry (1987), *The End of Organized Capitalism*, Cambridge: Polity Press.
Leroy, P. en J. van Tatenhove (2000), Political Modernization Theory and Environmental Politics, Buttel, F., G. Spaargaren and A.P.J. Mol eds., *Environment and Global Modernities*, London: Sage.
Marcuse, H. (1964), *One Dimensional Man. The Ideology of industrial Society*, London: Shere Books.
Mommaas, H. (1993), *Moderniteit, vrijetijd en de stad. Sporen van maatschappelijke transformatie en continuïteit*, Utrecht: Jan van Arkel.
Offe, C. (1986), Nieuwe sociale bewegingen als meta-politieke uitdaging, I. Weijers, eds., *Tegenspraken, dilemma's en impasses van de verzorgingsstaat, deel 2*, Amsterdam: SOMSO.
Ray, L.R. (1987), *Global Politics*, Boston: Houghton Mifflin.
Reinalda, B. (1997), *Private in Form and Public in Purpose: (I)NGO's as Political Actors in World Politics*, ECPR 25th joint sessions of workshops, Bern, Switzerland.
Risse-Kappen, T., eds. (1995), *Bringing Transnational Relations Back In. Non-State Actors, Domestic Structures and International Institutions*, Cambridge: Cambridge University Press.
Robertson, R. (1992), *Globalization. Social Theory and Global Culture*, London, Newbury Park, New Delhi: Sage.
Sachs, I. (1977), Het menselijke leefmilieu, J. Tinbergen, eds., *Naar een Rechtvaardige Internationale Orde. Een rapport van de Club van Rome*, Amsterdam: Elsevier, pp.309-321.
Schnaiberg, A. (1980), *The Environment from Surplus to Scarcity*, New York/Oxford: Oxford University Press.
Ullrich, O. (1979), *Weltniveau*, Berlin: Rotbuch Verlag.
Wagner, P. (1994), *A Sociology of Modernity. Liberty and Discipline*, London and New York: Routledge.
Wapner, P. (1995), Politics Beyond the State. Environmental Activism and World Civic Politics, *World Politics*, 47(3), pp. 311-340.
Williamson, P.J. (1989), *Corporatism in Perspective. An Introductionary Guide to Corporatist Theory*, London: SAGE.
Woods, L. (1993), Nongovernmental Organizations and the United Nations System: Reflection upon the Earth Summit, International Studies Notes, 1993(1), pp. 9-15.

3. Policy Arrangements

BAS ARTS, JAN VAN TATENHOVE AND PIETER LEROY

3.1 INTRODUCTION

In chapter 1 we defined the concept of institutionalisation as a process of structuration and stabilisation by which policy arrangements are produced, reproduced or transformed. This process was, moreover, analysed in terms of 'duality of structure', which implies that change will occur through agents-in-interaction as well as through long-term structural transformations. These latter transformations were, as far as the political domain of society is concerned, conceptualised in the previous chapter as political modernisation. We consider this to be a long-term background process, the three stages of which can be characterised by certain practices and discourses with regard to governance on the one hand and by state, market and civil society interrelations on the other. As stated earlier, these successive discourses and practices give rise to different 'ideal-types' of policy arrangements. Although obviously coloured by their period of birth, such policy arrangements are not entirely determined by a particular stage of political modernisation, as they are the outcome of specific interactions between certain actors in day-to-day policy processes. We consider policy arrangements therefore as the concept that links long-term processes of political change with specific processes of policy making and implementation on the ground.

In this chapter we consider the policy arrangement concept in depth. First of all we will define it in terms of the substance and organisation of a policy domain and elaborate on all elements of the definition. Secondly, we will operationalise the two aspects 'substance' and 'organisation' into four dimensions: 'policy coalitions', 'resources and power', 'rules of the game', and 'policy discourses'. Thirdly, we will elaborate on each of these four dimensions and deal with their interrelations. In addition empirical questions will be deduced from the theoretical analyses in each section, questions which are intended to guide the authors of the following chapters, although not all can or will be addressed in this book. Finally, we will again relate the concept of 'policy arrangement' to processes of institutionalisation, and argue that a plurality of arrangements does exist in every stage of political modernisation, linking traditional and innovative practices of policy making in any period of time. As a consequence, we put forward two types of analysis: a diachronic and a synchronic one, the former focusing on the plurality of arrangements over time (long-term), the latter on the plurality of arrangements in a specific, restricted period (short-term). In doing so, we believe that we have outlined a theoretical, heuristic framework, through which a non-evolutionary and non-instrumental analysis of (environmental) policy making and implementation can be made. Whether this belief is justified or not will be tested in the empirical chapters which follow.

3.2 DEFINITION AND OPERATIONALISATION

A 'policy arrangement' refers to the temporary stabilisation of the organisation and substance of a policy domain at a specific level of policy making. This definition requires explanation. Firstly, the concept of 'arrangement' was chosen to express our intention to focus more on 'stability' than voluntarist policy network analysts do (see the Introduction). In that sense, an arrangement links the interaction patterns of agents with the social-political structure within which these interactions take place. However, such a utilisation of the arrangement concept is not entirely new. Both sociologists and political scientists have already used it to refer to social and political systems, albeit without elaborating on it too much (De Swaan, 1988; Nelissen, 1992; Waltz, 1979).

Secondly, a policy domain in our perspective includes all policy practices with regard to an issue, such as climate change, nature conservation, acid rain, and the like. Although this is rather a broad description of a domain, it is a useful point of departure for any policy analysis. One way to delineate a domain is to focus (1) on the way in which a specific policy issue is formally institutionalised in a political system (ministries, departments, policy plans, etc.) and (2) on the way in which these formal boundaries are used and challenged by the relevant players in the field.

Thirdly, the notion of arrangement is defined above as stabilisation, which is not surprising given the day-to-day definition of an arrangement: 'putting things in *order*' (Oxford Advanced Learner's Dictionary). Although 'stabilisation' and 'order' are not synonyms, one may argue that as soon as a 'domain of things' exhibits some order, it also exhibits some stability. However, the reason for using the notion of stabilisation rather than order in our definition is twofold. The former has the connotation of process and, hence, possible change more than the latter does. Moreover, the notion of 'stabilisation' links our concept of arrangement to the literature on institutionalisation (see chapter 1).

Fourthly, the definition of a policy arrangement above refers to both organisation and substance, which we consider to be the two main *aspects* of an arrangement. Both are deduced from common sense views on policy practices. After all, organisation and substance are inherently linked in policy making and implementation. There is no such thing as a policy without *substance* (principles, objectives, measures etc.) and equally there is no such thing as a policy without *organisation* (departments, instruments, procedures, division of tasks and competence etc.). In addition, a time-space notion is part of the definition. On the one hand, any stabilisation of a policy domain is only temporary, as arrangements are under pressure through constant change, either by policy innovations on the ground or by processes of political modernisation. On the other hand, policy arrangements may evolve at different levels of policy making – local, national and international. Although the focus in this book will be mainly on the national level, some chapters will go beyond that (see chapters 5 and 6).

Having explained the definition of a policy arrangement, we need to elaborate on its main aspects – organisation and substance – in order to have a set of characteristics which will enable us to diagnose stability and change in policy arrangements. After all, these two aspects are still too broad and vague to guide policy analysis, and therefore have still to be operationalised (see figure 3.1). As far as *organisation* is concerned, we join Giddens' structuration theory and conceptualise organisations as social systems, being sets of agents that are nested in structures of rules and resources (Giddens, 1984). In this, three main dimensions of organisation can be distinguished: agents, rules and resources. These three – again – need to be elaborated on. As far as agency is concerned, the concept of 'policy coalition' is adopted here, as this is a popular unit of analysis at agent level in current policy sciences (Hajer, 1995; Sabatier, 1987). As players do not act individually in policy practices, but most of the time in *concert*, it is useful not to focus on individuals or on single organisations, but on (inter-organisational) coalitions. Exactly how will be dealt with in section 3.3. Resources, next, relate to power in organisation. Power is, according to us, an essential dimension of social and political life, and should therefore be part of any policy analysis. In section 3.4 this subject will be linked to policy arrangements. Rules, finally, co-determine how agents in organisations operate (thus reproducing and transforming those rules at the same time). We chose the description 'the rules of the game' in order to analyse this dimension in policy arrangements. This is a popular expression in political sciences to refer to procedures, conventions, and the like. Section 3.5 will cover this subject.

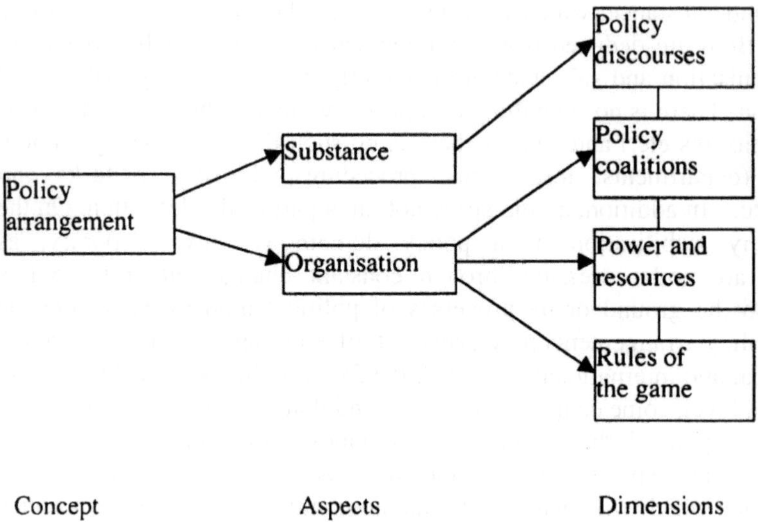

Figure 3.1: Operationalisation scheme of the concept of 'policy arrangement'

The second aspect of the policy arrangement concept, *substance*, is operationalised in terms of 'policy discourse'. Just as the notion of coalition is very popular in current policy sciences, so is 'discourse' (Dryzek, 1997; Hajer, 1995; Litfin, 1994). Generally, it refers to concepts, ideas, views, *buzzwords* and the like, which give meaning to a policy domain. It includes both formal policy plans and programmes and language practices which support or challenge these. Section 3.6 will deal with this fourth dimension of a policy arrangement.

Figure 3.1 summarises the path we followed in operationalising the concept of policy arrangement described above. It also shows that the four dimensions are interrelated (see the double arrows between the four boxes). For example, a policy discourse is (re)produced and transformed by the relevant policy agents, while at the same time this discourse co-determines which agents are legitimate stakeholders in policy practices. Equally, power relations co-determine which policy discourse may become dominant in a specific arrangement, whereas the rules of the game may exclude or include certain players. In other words, the four dimensions of a policy arrangement should be studied as one interrelated complex, in which distinctions have to be made for analytical reasons.

3.3 POLICY COALITIONS

In general a policy arrangement can be characterised by a (limited) number of coalitions. Each coalition consists of a number of players who share interpretations of a policy discourse or resources, in the context of the rules of the game (see below). As a consequence these coalitions identify (more or less) similar policy goals, and engage in policy processes to achieve those goals. In doing so, some coalitions may support the dominant policy discourse or rules of the game, while others might challenge these (*supporting* versus *challenging* coalitions). However, the formation and development of policy coalitions can be studied from at least two perspectives: from an institutional and from a strategic point of view.

Studying policy coalitions from the perspective of the actors' strategic conduct is comparable with the perspective of the policy network approach in policy sciences. Disregarding the numerous variants for a moment, this analysis focuses on the kind of social relations between interdependent actors which, primarily for strategic reasons, shape around policy problems and policy programmes in specific policy domains (cf. Kickert et al., 1997). Sharing a policy coalition is considered to be a strategic choice by actors, aiming to achieve their goals, and therefore looking for partners with whom policy interpretations are shared and an acceptable consensus can be reached. For example, in the Amsterdam Schiphol Airport case one can distinguish two or more opposing coalitions. The coalition in favour of unrestricted expansion of the airport (Schiphol, airline companies, the Department of Transport, employers' organisations, among others) is challenged by a coalition opposing such growth (the Green Left political party, the environmental movement, residents' groups, the Department of the Environment and others). How such coalitions are formed, which actors are involved or excluded, why some actors' strategic conduct succeeds while others' does not, and how and why some coalitions are successful, are some of the questions guiding the empirical research from this perspective.

The analysis of the formation and development of policy coalitions from an institutional perspective has its starting point in the process of political modernisation. In the phase of early political modernisation coalitions are likely to be formed within the nation state model or, at international level, as intergovernmental coalitions between nation states. In this way the early modernisation phase gave rise to more or less closed policy-making processes, dominated by small coalitions of state agents interacting with a small number of functional interest organisations. The direct input of (opposite) coalitions of for example interest groups, non-governmental organisations (NGOs), let alone non-organised citizens, was limited, as the main channel of political participation and public influence was considered to be general elections. Due to alternative ideas on governance and politics, resulting from the emergence of the anti-modernist wave of the 1970s, the coalitions broadened as various agents from both market and civil society got entry to (environmental) policy making and implementation. This broadening process became to some extent formalised in advisory and participation procedures: petition, lawsuit, impact assessment etc. This development enabled the formation of new policy coalitions, not only at national level, but also in international organisations such as the EU and UN. One should acknowledge, however, that the actual opportunities to intervene and the actual influence of those players and coalitions differed considerably.

As stated earlier, the late political modernisation stage has 'created' coalitions which cross the traditional divides and boundaries of the nation state model. Several academics (Beck, 1996; Wapner, 1995; Albrow, 1996 and Castells, 1996, 1997 and 1998) highlight the emergence of new coalitions. Beck's 'directed politics' for example refers to the constitution of a global civil society of *ad hoc* 'coalitions of opposites', which are in opposition to (early) modern political institutions. According to Castells (1996: 476): 'Dominant functions are organised in networks pertaining to a space of flows that links them up around the world, while fragmenting subordinate functions and people, in the multiple space of places, made of locales increasingly segregated and disconnected from each other.' From a post-modernist perspective, emphasising both the processes of globalisation and of individualisation, Albrow (1996: 121) states that 'modernity' is supplanted by 'globality'. That is a new level of organisation, to which any agent can relate, but which has no organising agent in itself.

All in all, the process of political modernisation is paralleled by the involvement of ever more private and public actors in national and international environmental politics, thereby inducing the broadening of policy coalitions. In addition, new and unexpected coalitions have recently been formed. An example is the co-operation of some business and environmental organisations, such as the insurance industry and Greenpeace, to achieve substantial reduction targets for greenhouse gases, attempting to reach their goals not only through intergovernmental politics but also through 'self-regulation' of market agents (Arts, 1998). The example makes clear the broadening of coalitions in two directions, as both horizontal interrelations (at one political level) and vertical interrelations (between different levels) have grown (Arts, 1990). International actors such as the World Bank and the global environmental movement collaborate with actors at national level, e.g. in forest policies in Brazil (Kolk, 1996) and in environmental politics in Mexico (Hoogenboom, 1998). At the same time, local farmers' organisations and local authorities take part in international policy making at UN headquarters. In other words, the local-global links have been intensified in policy coalitions recently (Princen and Finger, 1994). This process may be labelled the ***transnationalisation*** of policy coalitions (Hoogenboom, 1988).

The concept of policy coalition, and its numerous dimensions and aspects, raises interesting questions for empirical research. For example, what change in coalition building are we facing within environmental politics today? What kinds of coalitions are likely to result from different political contexts? How do the different policy interpretations of coalitions relate to the dominant policy discourse? Do the (supposed) increase in the number of actors involved and the broadening of coalitions go together with an increase in political participation, power and influence? To what extent and in which context are certain coalitions successful? These are, among others, the questions which guide the empirical chapters of this book.

3.4 RESOURCES AND POWER

The allocation and division of resources is a second important characteristic of policy arrangements, as these are 'systems of power'. However, power has – in accordance with 'the duality of structure' – to be regarded, on the one hand, as the ability of actors to mobilise resources and, on the other, as a relational and a structural phenomenon of social and political systems. The former refers to political power as a more or less permanent capacity of agents to maintain and transform their social or physical environment, and more specifically to achieve certain policy outcomes (Giddens, 1984; Guzzini, 1993; Held, 1995). Such outcomes may be achieved not only by determining political decisions, but also by dominating public debates, defining policy issues, setting agendas, or even changing the rules of the game (see next section), either at national or at international level (Bachrach and Baratz, 1962; Krasner, 1985; Strange, 1988).

However, to conceptualise power as mere capacity tends to ignore all those cases in which power is 'covertly', 'unnoticed' or 'unconsciously' exercised (Lukes, 1974). This means that the ability to achieve political outcomes also depends upon the relations of autonomy and dependency between actors 'in which these actors draw upon and reproduce structural properties of domination' (Giddens, 1981: 28-29). While it may be true that in the constitution of power agents are able to mobilise authoritative and allocative resources, it is also obvious that these resources are unequally divided among actors in any social system. In other words, power is about the asymmetrical distribution of resources (structural phenomenon), revealing itself in relations of autonomy and dependency between actors (relational phenomenon). The more these relations of power are objectified in institutional mechanisms and routines – fixing, so to speak, the allocation of competencies, qualifications, revenues and positions – the more natural and obvious domination seems (Frouws, 1993: 19).

Looking at the different stages of political modernisation, structures of domination change as a result of the changing interrelations between state, market and civil society. In the early and anti-modernisation stages, the state is the predominant 'power container'. Giddens reproduces the classical definition of the state when declaring that the nation state 'is a set of institutional forms of governance, maintaining an administrative monopoly over a territory with demarcated boundaries (borders), its rule being sanctioned by law, and direct control of the means of internal and external violence' (Giddens, 1985: 121). Yet one can distinguish different types of policy arrangements within this nation state model, implying different structures of domination. Corporatist arrangements, for example, can be characterised by a typical institutional form of exchange relations between the state and (a small number of) interest groups, to which a formal, representative status is granted in policy formulation, decision making and implementation. In this such interest groups are acknowledged as intermediate organisations, which implies a certain control over political resources, e.g. some formal authority. More specifically, the state provides information, entrance, influence, status and a quasi-monopoly, while the interest groups provide expertise, co-operation, discipline and legitimisation (Frouws, 1993: 43). This allocation of power to a large extent determines the capabilities of these actors (and of those not involved), and thereby the political outcome of this kind of arrangement.

In statist and liberal arrangements the allocation of authoritative resources and, therefore, the division of political power is different, leading to other kinds of resource dependencies (Liefferink, 1995). In liberal arrangements, for instance, state authorities are dependent on those who control resources they do not have themselves. For example, the need for economic expertise and investment capital makes state agencies dependent on companies, private consultants, banks etc., while in some cases NGOs such as Amnesty International or Greenpeace provide the necessary political legitimacy (Willets, 1982). It goes without saying that in such cases of mutual resource interdependency political power is less concentrated and more diffuse.

These concepts, again, are thought-provoking and raise questions for empirical research on policy arrangements in the environmental field. To name but a few: what does the allocation of resources and power look like in a specific environmental policy arrangement? How did this allocation change (or not) in different stages of political modernisation? What does the (assumed) broadening of policy coalitions mean for the allocation of power? What role do policy interpretations and discourses play in power processes? What do possible changes in formal power structures imply for the division of political influence? What happens in terms of power in those new policy arrangements which are based upon co-operative management, as these assume a more or less equal distribution of resources? Who takes over political power from the 'withdrawing state'?

3.5 RULES OF THE GAME

The rules of the game delineate a policy domain, i.e. they define the possibilities and constraints for policy agents to act within that domain. As such, these rules determine how politics is played, which norms are legitimate, and how policy outcomes are achieved, e.g. by which procedures, by which allocation of tasks, and by which division of competencies between actors and organisations. One can easily distinguish between formal and informal rules, the former being fixed and authorised in legal texts or documents, the latter being part of the predominant - but changing - political culture. A similar analytical distinction has to be made between rules of signification and legitimisation on the one hand, and rules used by actors in the policy-making process itself on the other.

In politics in general and in (re)producing and transforming policy arrangements in particular, actors constantly draw upon rules that provide them with guidelines to act properly and legitimately. In fact, these rules generate the definition of meaningful and justified circumstances. They delimit the agents who are the 'right' political players to be involved, and those who are not. In addition, they define the interrelations between actors, and, as a consequence, they delineate the boundaries of policy coalitions: who is in and who is out; how one can get in; what the relationship with outsiders is; etc. Moreover, these rules describe how the political game should be played: how issues may be raised; agendas set; interests articulated; policies formulated; decisions made; and measures implemented. In short, the rules of the game encompass all modes of production and interpretation of meaningful and legitimate conduct in (environmental) policy arrangements, implying the self-conscious application of normative and interpretative schemes (discourses) by the actors involved, which are included in sanctioning procedures (see Giddens, 1984; Cohen; 1989).

It is obvious that the rules of the political game change continuously over time, partly as a result of long-term societal processes such as political modernisation, partly caused by specific interventions of and interactions among actors in tangible policy-making processes. We indicated in the previous chapter how the theorists of the anti-modernisation thesis challenged the then dominant political culture, advocating a different political order, including new rules for the political game. Subsequently, protagonists of the post- or reflexive modernisation thesis explicitly defended the proposition that the rules of the game themselves have become the object of ongoing reflection (Beck, 1997). In their view, the rules are no longer accepted as pre-given, or as fixed within the nation state model, but are continuously challenged through political practices. One should remember here Beck's distinction between 'rule-directed' and 'rule-altering' politics, the former referring to politics within the nation state model, the latter to sub-politics in the global risk society. Following from this it is plausible to see traditional policy arrangements as being characterised by *rule-directed* politics, in which the rules reflect the early modern structures of signification and legitimisation within the context of specific interrelations between state, market and civil society. Although they differ considerably, this also applies to statist, corporatist and liberal policy arrangements. Policy arrangements exceeding this 'modern' scheme consist of broad and diffuse policy coalitions, as referred to by Beck, Castells, Albrow and others, and therefore *rule-altering* politics characterises these arrangements more adequately.

Although the idea of more fluid and less formalised rules of the game in the post- or reflexive modernisation era seems obvious at first sight, we again have to ask empirical questions regarding the actual contents of this change. To enumerate but a few: did the process of political modernisation bring about new rules regarding the entry to and the interactions within political arenas? And what do these new rules look like? What are the consequences, for NGOs for example, of getting entry to political arenas, at local, national and international levels? To what extent does co-operation with state agencies imply incorporation and co-optation in terms of taking over predominant discourses (see next section), existing rules of the game, the use of expertise, the disciplining of grassroots supporters, and the narrowing of strategic opportunities? To what extent are Beck and others too optimistic about the rule-altering opportunities in contemporary politics? What exactly are the rules of the game behind the 'wave' of new policy arrangements based upon consensual management and negotiated decision making? And how do these rules relate to the allocation of power we discussed earlier?

3.6 POLICY DISCOURSES

The transformation of environmental politics is not only affected by the process of political modernisation, by new governance models and by changing state-market-civil-society interrelations, but also by the way environmental problems are redefined socially and politically. This is done through *discourse*. To understand and give meaning to environmental problems and to design possible solution strategies, actors draw upon interpretative schemes or environmental policy discourses. A policy discourse can be defined as: 'A specific ensemble of ideas, concepts, and categorisations that are produced, reproduced and transformed in a particular set of practices and through which meaning is given to physical and social realities' (Hajer, 1997: 44). Dryzek defines a policy discourse similarly as: 'A shared way of apprehending the world. Embedded in language, it enables those who subscribe to it to interpret bits of information and put them together into coherent stories or accounts. Each discourse rests on assumptions, judgements and contentions that provide the basic terms for analysis, debates, agreements and disagreements (...)' (Dryzek, 1997: 8).

On the basis of these two definitions and in accordance with structuration theory, we define policy discourses as dominant interpretative schemes, ranging from formal policy concepts to popular story lines, by which meaning is given to a policy domain. In general, a policy arrangement can be characterised by one dominant policy discourse, the content of which is continuously challenged by (elements of) competing discourses.

Examples of policy discourses in the environmental field are 'small is beautiful', 'eco-development', 'political ecology', 'ecological modernisation' and 'sustainable development' (respectively Schumacher, 1974; Sachs, 1977; Benton, 1997; Huber, 1985; Mol, 1995 and WCED, 1987). Although some of these notions refer to detailed theories and programmes, others are no more than popular *buzzwords*. Nonetheless, they all stand for (explicit or implicit) interpretative schemes by which meaning is given to environmental problems, and on the basis of which environmental policy programmes are designed. Political ecologists, for example, believe that the ecological crisis is due to the capitalist nature of our economy and they therefore strive for a green-socialist society, whereas the programme of ecological modernisation foresees a solution of the ecological crisis within the capitalist economy through technological innovation. Similarly, supporters of the discourses 'small is beautiful' and 'eco-development' promote a small-scale society, both in the North and the South, whereas 'sustainable development' advocates a large-scale, resource- and energy-efficient economy.

Consistency, elaboration, operationalisation and coherence, however, are not essential features of discourses (Hajer, 1997). This becomes obvious from the 'sustainable development' example, which is interpreted quite differently by various actors (Arts, 1994). One can even argue that the vaguer a policy discourse is, the more open it is to different interpretations, the greater its mobilising capability, and the more impressive its consensus-building ability is. Story lines, being the attractive and often seductive one-liner versions of policy discourses, may therefore enable various or even opposing policy actors to join forces. In that respect 'sustainable development' has been very successful *hitherto* at international level, while 'ecological modernisation' is the successful catchword of environmental policies in various Western nations.

The changes in the predominant environmental discourses of the 1970s, 1980s and 1990s are remarkable (cf. Dryzek, 1997). The anti-modernisation wave of the 1970s was to a large extent responsible for the rise of environmental issues. The dominant discourses ('political ecology', 'eco-development', 'small is beautiful', and others) were anti-statist and anti-capitalist in nature, causing considerable gaps between mainstream politics and the countervailing power of, among others, the emerging environmental movement. The latter, incidentally, defined itself as a *counter*-movement, and phrased environmental issues primarily in terms of conflict. The late 1980s and 1990s in contrast are characterised by all-embracing consensus discourses ('ecological modernisation', 'sustainable development'), aimed at and to a large extent succeeding in getting support from right, left and green alike. In these contemporary discourses environmental problems are seen as manageable issues, to be managed not only by the state, but also by a wide range of co-operative stakeholders. In addition, problems are conceived as challenges to innovation – technological rather than political – and as opportunities to reconcile economy and ecology – therefore urging state, market and civil society agents to contribute their share. From this it becomes clear that these (successful) environmental policy discourses combine elements of the political modernisation thesis, at least those that are specific to environmental problems and policies, and include an appeal for new coalitions and new rules of the game. The above shows that environmental policy discourses, albeit implicitly, contain more general statements and positions on politics and policies, on the role of the state, and the like.

These recently developed 'consensus' discourses seem to provide the legitimisation for the new environmental policies, which have resulted in a variety of innovative arrangements, and which reflect the thesis of political and ecological modernisation to a very large extent. Yet a great many empirical questions are still to be answered. What exactly is the role of policy discourses in the renewal of environmental policies and in the broadening of policy coalitions? What are the typical 'biases' and 'reductions' of the contemporary consensus discourses? In other words, which issues are defined into politics and which are defined out by these discourses (Schattschneider, 1960; Hajer, 1997)? Which opportunities do they stress and which uncertainties and risks do they tend to neglect? How and by whom are these discourses (re)produced and renewed? What are the roles of politicians, academics, private consultants and other actors in the launching and use of policy discourses? And do the 'old radical greens' have a future, in view of the overwhelming success of consensual story lines? These and related questions will guide us in analysing the recent changes in specific environmental policy domains in depth.

3.7 PLURALITY: DIACHRONIC AND SYNCHRONIC PERSPECTIVES

In the previous chapter as well as above we have analysed the process of political modernisation, its successive stages and their impacts on political practices, especially on the (re)production and transformation of environmental policy arrangements, resulting in the rise and fall of certain policy coalitions, power relations, rules of the game and discourses. In doing so, we might have suggested that we support an evolutionary model of change, from early via anti- to late modernisation. Although certain periods in the overall process of political modernisation are likely to bring about specific ideal-types of policy arrangements, this does not imply a pre-given, evolutionary and linear modernisation of both politics and arrangements. In passing, we have already focused on a range of arrangements in each stage. For example, the stage of early modernisation was characterised by both statist and corporatist arrangements. Similarly, the stage we are in today can be characterised by both liberal and reflexive arrangements, and – very likely – by hybrids of these two as well. Therefore we like to speak of a *plurality* of policy arrangements.

This plurality can be studied from at least two angles. The *diachronic* perspective highlights changes in terms of policy coalitions, resources and power, rules of the game and policy discourses during the course of history, extracting the empirical evidence from the rise and fall of certain types of policy arrangements over time. At the same time, though, as none of the successive stages exclusively brought forward arrangements of a single type, we do need a *synchronic* perspective as well. In this perspective the co-existence of different types of arrangements in every period is highlighted, providing a more sophisticated outlook upon the plurality of policy arrangements in different subdomains of environmental politics. By emphasising this plurality from two perspectives, we envisage preventing a too simplistic, evolutionary view of policy change.

3.8 CONCLUSION

In this chapter we introduced the concept of (environmental) policy arrangement. An arrangement can be characterised by: (1) the nature of the policy coalitions involved (their number, quality etc.); (2) the division of resources within and between coalitions and the allocation of political power; (3) the predominant policy discourses and the contents of the political debate; and (4) the nature of the rules of the political game. Together with the political modernisation thesis, outlined in chapter 2, this policy arrangement concept is needed to construct our argument, namely that both organisation and substance, both structural and agency variables, both long-term and short-term processes are crucial phenomena in understanding stability and change in environmental politics. Moreover, we added the 'diachronic' and 'synchronic' analyses of the plurality of policy arrangements in order to overcome a mere evolutionary perspective on policy change. In the empirical chapters 4 to 8, this theoretical framework will be applied in detail to several environmental domains at several levels of policy making.

REFERENCES

Albrow, M. (1996), *The Global Age. State and Society Beyond Modernity*, Oxford: Polity Press.

Arts, B. (1990), Brundtland en het concept van interdependentie, B. Arts, eds., *Ontwikkeling, Milieu en Veiligheid: Mondiale Interdependentie*, Amsterdam: Uitgeverij Jan Mets, pp. 5-35.

Arts, B. (1994), Dauerhafte Entwicklung: eine begriffliche Abgrenzung, *Peripherie. Zeitschrift fur Politik und Okonomie in der Dritten Welt*, 14(54), pp. 6-27.

Arts, B. (1998), *The Political Influence of Global NGOs. Case Studies on the Climate and Biodiversity Conventions*, Utrecht: Van Arkel/International Books.

Bachrach, P. and M. Baratz (1962), Two Faces of Power, *American Political Science Review*, 56, pp. 947-952.

Beck, U. (1996), World Risk Society as Cosmopolitan Society? Ecological Questions in a Framework of Manufactured Uncertainties, *Theory, Culture & Society*, 13(4), pp. 1-32.

Beck, U. (1997), Global Risk Politics, M. Jacobs, eds., *Greening the Millennium? The new politics of the environment*, Oxford: The Political Quarterly/Blackwell, pp. 18-33.

Benton, T. (1997), Beyond Left and Right? Ecological Politics, Capitalism and Modernity, M. Jacobs, eds., *Greening the Millennium? The new politics of the environment*, Oxford: Blackwell, pp. 34-46.

Castells, M. (1996), *The Information Age: Economy, Society and Culture, Volume I, The Rise of the Network Society*, Oxford UK, Malden USA: Blackwell.

Castells, M. (1997), *The Information Age: Economy, Society and Culture, Volume II, The Power of Identity*, Oxford UK, Malden USA: Blackwell.

Castells, M. (1998), *The Information Age: Economy, Society and Culture, Volume III End of Millennium*, Oxford UK, Malden USA: Blackwell.

Cohen, I.J. (1989), *Structuration Theory. Anthony Giddens and the Constitution of Social Life*, London: MacMillan Education Ltd.

De Swaan, A. (1988), *In care of the state: health care, education and welfare in Europe and the USA in the modern era*, Cambridge: Polity Press.

Dryzek, J.S. (1997), *The Politics of the Earth. Environmental Discourses*, Oxford: Oxford University Press.

Frouws, J. (1993), *Mest en macht. Een politiek-sociologische studie naar belangenbehartiging en beleidsvorming inzake de mestproblematiek in Nederland vanaf 1970*, Wageningen.

Giddens, A. (1981), *A Contemporary Critique of Historical Materialism. Vol. 1 Power, property and the state*, London, Basingstoke: The MacMillan Press Ltd.

Giddens, A. (1984), *The Constitution of Society. Outline of the Theory of Structuration*, Cambridge: Polity Press.

Giddens, A. (1985), *The Nation-State and Violence*, Cambridge: Polity Press.

Guzzini, S. (1993), Structural power: the limits of neorealist power analysis, *International Organization*, 47(3), pp. 443-478.

Hajer, M.A. (1997), *The Politics of Environmental Discourse: Ecological Modernization and the Policy Process*, Oxford: Oxford University Press.

Held, D. (1995), *Democracy and the Global Order. From the Modern State to Cosmopolitan Governance*, Cambridge: Polity Press.

Hoogenboom, B. (1998), *Mexico and the NAFTA Environment Debate. The Transnational Politics of Economic Integration*, Utrecht: Van Arkel/International Books.

Huber, J. (1985), Ecologische modernisering, E. Van den Abbeele, eds., *Ontmanteling van de groei. Leesboek over een andere economie*, Nijmegen: Markant, pp. 161-168.

Kickert, W.J.M., E.H. Klijn and J.F.M. Koppenjan eds. (1997), *Managing Complex Networks. Strategies for the Public Sector*, London: SAGE.

Kolk, J.E.M. (1996), *Forests in international environmental politics: international organisations, NGOs and the Brazilian Amazon*, Utrecht: International Books.

Krasner, S.D. (1985), *Structural conflict: The Third World against global liberalism*, Berkeley: University of California Press.

Liefferink, J.D. (1995), *Environmental policy on the way to Brussels. The issue of acidification between the Netherlands and the European Community*, Wageningen: WAU.

Litfin, K.T. (1994), *Ozone Discourses: science and politics in global environmental cooperation*, New York: Columbia University Press.

Lukes, S. (1974), *Power. A Radical View*, London: Macmillan.

Mol, A.P.J. (1995), *The Refinement of Production. Ecological modernization theory and the chemical industry*, Utrecht: Van Arkel.

Nelissen, N.J.M. (1992), *Besturen binnen verschuivende grenzen*, Zeist: Kerckebosch.

Princen, T. and M. Finger (1994), *Environmental NGOs in world politics. Linking the local to the global*, London: Routledge.

Sabatier, P.A. (1987), Knowledge, Policy-Oriented Learning, and Policy Change, and advocacy coalition framework, *Knowledge: Creation, Diffusion, Utilization*, 8(4), pp. 649-692.

Sachs, I. (1977), Het menselijke leefmilieu, J. Tinbergen, eds., *Naar een Rechtvaardige Internationale Orde. Een rapport van de Club van Rome*, Amsterdam: Elsevier: pp. 309-321.

Schattschneider, E.E. (1960), *The semisovereign people: a realist's view of democracy in America*, New York: Holt, Rinehart and Winston.

Schumacher, E.F. (1974), *Small is Beautiful. A Study of Economics as if People Mattered*, London: Abacus.

Strange, S. (1988), *States and Markets*, London: Pinter.

Waltz, K.N. (1979), *Theory of International Politics*, New York: McCraw-Hill.

Wapner, P. (1995), Politics Beyond the State. Environmental Activism and World Civic Politics, *World Politics*, 47(3), pp. 311-340.

Willets, P., eds. (1982), *Pressure Groups in the Global System*, London: Frances Pinter.

World Commission on Environment and Development (1987), *Our common future*, Oxford: Oxford University Press.

4. Dutch Infrastructure Policies: Changing and Contradictory Policy Arrangements

PAUL PESTMAN

4.1 INTRODUCTION

In this chapter postwar changes in Dutch policy arrangements in the field of transport and infrastructure will be analysed, by considering both the dominant form of policy making and the dominant form of discourse in a given period. Four phases are identified, three of which will be described in section 4.2. The main part of the chapter focuses on recent changes in Dutch infrastructure policies. Since the beginning of the 1990s a number of institutional changes have taken place. Policy making has become more transparent, new actors have been invited to participate and more attention has been paid to communication. At the same time there has been a substantial increase in the resources available for infrastructure and a strong desire to implement a large number of new infrastructure projects as soon as possible. This desire reduces the scope for public participation. These contradictory transformations will be described in section 4.3. In section 4.4 the consequences of these changes will be illustrated by examining three recent policy initiatives. Section 4.5 contains some concluding remarks.

4.2 DUTCH POLICY ARRANGEMENTS FOR TRANSPORT AND INFRASTRUCTURE 1945-1989: GROWTH, POLITICISATION AND DECLINE

4.2.1 The concept of policy arrangements

A policy arrangement is defined as a temporary stabilisation of the organisation and content of a policy domain. The organisation of a policy arrangement is characterised by the actors taking part, the existing political rules and the distribution of resources. In the analyses of policy arrangements differences in centralised/decentralised policy making, public participation, openness, and state-interest group relations become visible. The prevailing discourses in a policy domain form the second of the elements which make up a policy arrangement. According to Hajer, a discourse is a specific ensemble of ideas, concepts and categorisations which are produced, reproduced and transformed in a particular set of practices and through which meaning is given to physical and social relations (Hajer, 1995: 44). Issues such as what infrastructure policies should be about, which important aspects should be taken into account and who is responsible for it are part of this second element. Together, organisation and content of a policy domain form the characteristics of a policy arrangement at a given point in time. Several policy arrangements may be found in a policy field; demarcation of specific policy arrangements takes place on the basis of an observed degree of coherence in organisation and content.

Policy making is relative unproblematic when there is a conformity between the policy issue at hand and the existing coalitions of actors (conformity of interests), the rules (institutional conformity), and the resources (material conditions). It is easy to put a new policy issue on the political agenda when there is such conformity, that is if the issue in question is in keeping with the existing policy arrangement. In the absence of such conformity, political agenda setting is dependent on the introduction of a new discourse, through which actors are able to redefine their interests and the possibilities of forming coalitions. Discourses should therefore not only be seen as a reflection of cognitive patterns of actors. As a result of a new discourse social reality becomes redefined. This phenomenon can be called *performative* mobilisation. This is one of the ways in which the changing of policy arrangements can be explained.

4.2.2 Reconstruction and growth (1945-1971)

In the early postwar period infrastructure was of great importance. The damage caused by the Second World War was considerable. Reconstruction of the Dutch economy was the primary policy goal and the need for investment in infrastructure was not in doubt: there was national consensus on this subject.

Traditionally, the Ministry of Transport, Public Works and Water Management (*Ministerie voor Verkeer & Waterstaat* - V&W) has been the main policy actor in the infrastructure field. It keeps relatively strong control over the policy-making process. The Ministry has a long-established engineering department, *Rijkswaterstaat*, which expends most of the Ministry's human and financial resources. Rijkswaterstaat has a high level of technical expertise and is implementation-oriented; in principle problems are considered to be solvable and manageable. Rijkswaterstaat has generally used a somewhat technocratic working method (De Heer, 1991).

In deciding on the siting of new roads, Rijkswaterstaat followed the idea of 'sectoral corporatism', which in the 1950s was the dominant mode of policy making, in which specific interest groups were involved in the policy-making process. Advisory councils were formed in all kinds of policy fields (Van Delden, 1985: 150-151), and the Water Management Council (*Raad voor de Waterstaat*) was established in 1950. An important committee of this council was the Consultative Committee on Roads (*Commissie Overleg Wegen* - COW), on which administrators, interest groups and independent experts sat.

Although the parliament feared that the Council would weaken its own role (*Handelingen Tweede Kamer 1949-1950 1315 nr. 4*), the COW had virtually no influence either on policy making or on the siting of specific roads in this period (Huberts, 1988: 251). Rijkswaterstaat prepared plans for new roads and other infrastructure and the committee advised on these plans. The COW was willing to listen to the department's suggestions. Project plans and alternative routes were not explored in any detail. Only the route preferred by Rijkswaterstaat was given attention (Huberts, 1988: 251). The nature conservation groups, organised in the *contact commissie*, largely assented to the Ministry's plans and made minor submissions only.

4.2.3 Politicisation (1971-1982)

In the 1970s both the dominant form of policy making and the dominant policy discourse were challenged. To start with the policy making, an increasingly wealthy and well-educated population demanded a more transparent style of decision making. The Minister of Housing and Spatial Planning (VRO) had already in June 1968 asked the Council for Spatial Planning (Raro) for advice on how to involve the public in the development of spatial policy (Raro, 1970a). This council concluded in the case of the siting of a Shell refinery at Moerdijk in the southwest of the Netherlands that 'the government did not take all relevant interests into account in a clear way. This could be done by means of a special planning procedure' (cited in Van der Cammen and De Klerk, 1993: 175). The report that ensued forms the basis for the current procedure for making key decisions on spatial planning, which was formalised in 1985 (Raro, 1970b).

Not only was the organisation of the policy field being contested, the consensus on the construction of roads that had existed in the 1950s and 1960s also broke up. Several political decisions, such as those on the extension of the road network and the specific location of routes, were forcefully challenged (see Huberts, 1988: 250-255). New environmental groups were formed and became new actors in the field. They were much more critical with regard to new infrastructure and as a result environmental problems became a dominant feature of public debate.

In the province of Brabant the construction of the A75 motorway near Eindhoven was disputed, as well as that of the A2/264. The Ministry was forced to cope with action taken outside the parliament. The political rules for policy making in this field had to be changed considerably as a result and a new way of legitimisation had to be established. In this we can recognise elements of the phase of political anti-modernisation, discussed in chapter 2. There were outbursts of radical protest and the Ministry had to deal with a polarised and politicised consultation arena. The newly-emerging activism led to the reshaping of the nature conservation and environmental protection organisations, culminating in the Organisation for Nature and Environment (*Stichting Natuur en Milieu* - SNM). Because SNM replaced the contact commission as the environmental representative on the COW, this new movement also gained institutional access to the field.

The Ministry reacted to these political and societal changes by reflecting on its own position and working methods. On 31 July 1970 the Minister of Transport announced that in addition to legal procedures already in the pipeline, the Ministry intended to start a change in practices in the policy preparation phase. Experiments were to be conducted with giving citizens the opportunity to contribute to the development of plans for main roads (Raro, 1970b: 1). As a result of these new procedures the COW became more influential, but also more divided. The COW itself also changed its working practices. Until the 1970s only formal institutions were allowed to participate in a hearing, but from then on everyone was entitled to do so.

As a result of the growing criticism of unrestrained growth in car use a number of road-building plans were cancelled. Although the regional departments opposed this policy change, the hierarchical relationship between them and the Ministry made it possible (Huberts, 1988: 255). Furthermore, the Ministry increased the number of possible alternative routes which should be studied when deciding on the construction of new roads. Research activities were extended and there was a growing awareness within the Ministry that decision making needs multilateral reasoning, that integrated management is needed, and that policy evaluation is an essential element of the policy cycle. During the preparation and implementation of water engineering projects, for example, attention was now paid to safegarding ecological values and to landscaping. This meant that environmental knowledge became a resource for effective policy making in the infrastructure field. The first example of this was the Deltaworks, intended to prevent disasters such as the floods of February 1953, when half the province of Zeeland was under water and 1,835 people drowned. The masterpiece was the construction of the Oosterscheldedam, which was completed in 1987. Its construction allows the tidal system in the Oosterschelde to be maintained, while protecting the area from floods during storms and spring tides. The dam can be viewed as a successful, though expensive, compromise between security against flooding and maintaining the tidal regime and its accompanying wildlife. In general, the words 'defence' and 'response' (De Heer, 1991: 149-150) sum up the dominant mode of policy making in the 1980s. At the beginning of the 1980s the research efforts were further increased. Existing expertise in environmental affairs was combined and extended, partly through the reorganisation of the environmental services, when the Deltaworks were in the final stages of construction (De Heer, 1991: 160-161).

In 1981 the first Integrated Transport and Traffic Plan (*Structuurschema Verkeer en Vervoer - SVV*) was published. Its purpose was to give information on the long-term policy plans for transport and infrastructure and their spatial consequences (1981: 3). SVV exhibits a shift from the construction-oriented approach to transport and infrastructure towards a more balanced approach. It can be seen as foreshadowing a change in the policy discourse.

4.2.4 Decline (1982-1989)

At the beginning of the 1980s the Dutch economy was in a bad state. Mass unemployment and decreasing state revenues from natural gas dominated public debate. The government reacted with severe budget cuts, budget discipline and privatisation in order to solve its own financial problems. During the 1980s doubts were expressed about state intervention, as was also the case in other countries.

Transport policies in this period showed three features. First, large policy innovations were absent, not much was happening. Existing road construction projects were delayed and both financial resources and political will for innovative plans were simply absent. During the 1980s the political context was one of budget cuts. Secondly, public participation in decision making turned out to be more difficult than had been expected in the 1970s. Administrative consensus was reached before the public was invited to take part in the process, which led to a situation where citizens felt that every plan had been fixed beforehand, and that participation did not serve any useful purpose. As a result citizens began to use available legal instruments to obstruct decision-making processes (Veldboer, 1996). Thirdly, problems caused by the increase in mobility, such as traffic jams and air pollution, were seen as growing and demanding a policy response. Rijkswaterstaat became conscious of the negative side effects of its building programmes.

By the end of the 1980s the field of transport and infrastructure had its own specific developments (budget cuts and the completion of the Deltaworks). However, the existing policy arrangements had been influenced by elements of early political modernisation as well as political anti-modernisation. The Ministry of Transport remained the main policy actor, occupying a strong institutional position. The methods of policy making remained somewhat technocratic, but to a much greater extent it was now society which defined policy problems. Environmental knowledge and skills had become crucial in addition to the engineering skills. The environmental movement had gained institutional access to the policy-making arena. The consensus of the period after the war had been replaced by several contrasting views on spatial and economic development. The Ministry of Transport reflected this situation by introducing integrated policy plans (*Structuurschema's - SVVs*) to cope with problems of congestion, safety and comfort. Contact with the public was largely through formal participation, a system that was clearly not functioning well.

In recent years there has been a large shift in infrastructure arrangements. These changes will be analysed in the next section. However, in order to give a full overview of the changes in the postwar period, the main characteristics of this renewal are included in table 4.1.

	Societal development	Dominant form of policy making	Dominant policy discourse
1945-1971	Consensus over main economic goals	Technocratic policy making with formal corporatist elements	Reconstruction of the Dutch economy
1971-1982	Transparency, polarisation	Introduction of public participation in the technocratic scheme	Multilateral decision making - reduction in road construction and 'landscaping'
1982-1989	Economic decline, budget cuts	Decrease of public participation	No money or political will for new plans
1989-2000	Globalisation, European integration and economic prosperity	Centralisation on the one hand and interactive planning on the other	Mainport development and search for public support

Table 4.1: Characteristics of policy arrangements in infrastructure policies

4.3 NEW, CONTRADICTORY POLICY ARRANGEMENTS (1989-2000)

At the end of the 1980s infrastructure policies faced difficult times. Public participation did not function properly and the economic crisis made it impossible to develop new policy plans. However, developments in two other policy fields influenced the position of infrastructure policies dramatically. Both the National Spatial Planning Agency (RPD) and the Ministry of Economic Affairs were reconsidering their strategic goals. Motivations and backgrounds for these reorientations were different, but the result of it was that the policy fields of the Ministry of Economic Affairs, the Ministry of Transport and the RPD became more closely linked. As result a new coalition emerged that focused on the renewal of the spatial-economic structure. Infrastructure was seen as the motor for the national economy and legal and financial instruments were designed for the implementation of these plans. At the same time new arrangements for public participation were formulated as well. In following sections these changes will be described, starting with the reorientation of important policy actors.

4.3.1 Policy reorientations in spatial planning and economic affairs

Spatial planning looking for a new goal

Driven by the report of the Club of Rome, the central policy goals of economic policy were reconsidered. In 1976 the National Spatial Planning Agency (RPD) published the Third Document on Spatial Planning. Its main objective was to promote spatial and ecological conditions which would benefit the essential requirements of individuals and groups, while diversity, cohesion and sustainability of the environment was guaranteed (Ministerie van VRO, 1973: 99). Spatial policies have a threefold purpose:

- The solution and prevention of spatial conflicts,
- The promotion of societal needs (comfort and regional economic conditions, the regional dispersal of economic development),
- The improvement of living conditions (Ministerie van VRO, 1973: 102-104).

These goals reflect the dominant political and societal trends of the 1970s. The main focus of the document was the strengthening of weak regions and the protection of vulnerable aspects of society. By 1982 the policy concepts of the Third Document were seen as outdated. The RPD's right of existence was challenged and within the RPD a strong need was felt to rethink spatial problems. The main question the RPD posed itself was which issues for the 21st century could be solved with the help of spatial policies? One of the projects the RPD initiated was 'Spatial explorations for infrastructure' (*Ruimtelijke Verkenningen Infrastructuur* - RUVEIN). The study was published in 1985 and its main conclusion was that while the existing transport connections had a national character, due to European integration and other international developments the Netherlands should start to think in terms of international connections. One of the points made was that the construction of new railway lines for freight should be considered. The port of Rotterdam and Amsterdam's Schiphol airport were seen as the most important areas for economic development.

In 1988 the Fourth Policy Document on Spatial Planning (*Vierde nota inzake de Ruimtelijke Ordening* - VINO) was published. Compared to the Third Policy Document it shows a strong shift in emphasis from the weak to the strong. Instead of the dispersal of economic activities, centralisation was promoted, and the heavy bias towards the protection of weak sectors of society was replaced by spatial planning as an instrument to achieve the main (economic) national goals of economic growth and full employment. 'Mainports' and 'transport axes' became central concepts.

Economic policy: from the weak to the strong

In the 1970s Dutch economic policy was characterised by dispersion and selective growth and facet policy (Ministerie van Economische Zaken, 1976; see also Knoester, 1989: 136). Economic activities had to be dispersed equally throughout society. Governmental institutions were moved to areas with a marginal economic position and subsidies were made available to direct economic activities (*Wet Investeringsrekening, Wet Selectieve Investeringsregeling, Regionale investeringspremieregeling*) (Knoester, 1989: 136). In 1972 the Dutch Company for Reconstruction (*Nederlandse Herstructureringsmaatschappij*, NEHEM) was established.

As a result of the oil crisis the focal point of structural economic policy moved from generic to specific policies. At the end of the 1970s companies which were not able to compete with international competitors received financial support and during the early 1980s this support was extended (Knoester, 1989: 154).

In 1981 the advisory commission for industrial policy (Commission Wagner), promoted a new industrial vigour. The commission's report led to the establishment of the Company for Industrial Projects (*Maatschappij voor industriële projecten - MIP*). This type of policy making bore a strong resemblance to the politics of industrialisation of the postwar period, as promoted by the Minister of Economic Affairs Van den Brink (Knoester, 1989). It marked a change from support for individual companies to sectoral structural policy. Furthermore, technological innovation became the buzzword in economic development. In general it also meant a move away from 'loser' towards 'winner'. The public sector faced decreasing budgets and privatisation.

By the end of the 1980s two developments became important: globalisation and European integration. The new direction within the Ministry of Economic Affairs can be seen in the note 'An economy without frontiers' (*Economie zonder grenzen*) (Ministerie van Economische Zaken, 1990), which focused on international trade and transport axes, supported by among others the port of Rotterdam. Later, the importance of infrastructure and spatial planning to economic development was conceptualised in 'Space for regions' (*Ruimte voor regio's*) (Ministerie van Economische Zaken, 1995) and 'Space for economic dynamics' (*Ruimte voor economische dynamiek*) (Ministerie van Economische Zaken, 1997). After the withdrawal of the interventionist state, the state was now brought back in, but in a specific form: intervention through spatial economic planning and investment in infrastructure.

4.3.2 Renewal of transport policies: the emergence of the 'mainport' discourse

During the 1980s declining budgets shaped the political context for all ministries, including the Ministry of Transport. Rijkswaterstaat even had problems with the financing of road maintenance. The Deltaworks, Rijkswaterstaat's main activity for many years, had been completed. The report 'Spatial explorations for infrastructure' by the National Land Use Planning Agency made an impact at the Ministry of Transport. They were not informed about the study in advance, which caused some discontent, but quite soon this turned into support, because the study was orientated towards the future and opened up new possibilities for the Ministry. As a result of the policy reorientations transport and infrastructure gained a new position on the political agenda.

At the same time the Ministry became aware of the negative side effects of its building programmes. In the second Transport and Traffic Plan (SVV 2) the Ministry promoted a transport system 'which does not transfer environmental problems to future generations. A sustainable society contains an intelligent and well-thought-through system of transport connections, with limits on the extension of infrastructure. Only truly necessary incursions into the landscape will be tolerated' (Ministerie van Verkeer en Waterstaat, 1990: 8). SVV 2 was the product of two different discourses. The first is that on comfort. However, later in the report it is stated that transport policies are designed to ensure good accessibility. 'The maintenance and strengthening of infrastructure, to attract international corporations, is an important goal of the government. This requires that the driving forces of our economy - the mainports Schiphol and Rotterdam - have good road, water, railway and information technology connections'. The tension between these two aspects can be felt throughout the report. Nevertheless, one key issue in which both were supposed to be able to meet was the construction of new railwaylines, as this appeared to be a mode of transport which caused less environmental damage than others. This notion formed the starting point of the ongoing renewal of the spatial-economic structure taking place during the 1990s (Betuweroute and High Speed Rail).

We can observe here the rapid progress of a discourse on spatial development and transport which can be called the 'mainport' discourse. This discourse tells us something about the main policy goals actors find attractive, the international trends and developments which are seen as important, the role of the state in response to this and the preferences for a specific way of dealing with conflicts and questions of decision making.

The 'mainport' discourse consists of three elements. The first is the idea that economic development in the Netherlands can be best promoted by concentrating economic activities in the so-called 'mainports'. The term 'mainport', which is not a recognised word in the English language, became extremely popular in Dutch debates on economic development. Mainports are very large seaports or airports, which have a significant international position in terms of both quantity and quality and are assumed to have great impact on a country's national economy. Mainports are located at the junction of a number of transport axes and should therefore have adequate infrastructure available for several modes of transport (road, water, rail, pipeline). Transport is seen as a major driving force for economic development, which places the construction of infrastructure and spatial planning at the centre of economic policy making. The port of Rotterdam is considered to be a mainport, although in the past its emphasis was on quantity rather than quality. Amsterdam's Schiphol airport is seen as an economic area which can be developed further, into a mainport (VINEX, VINO, SVV2, PKB, and Note *Versterking Ruimtelijk Economische Structuur*). There are a number of recurring trends and developments that are strongly connected with the idea of mainports. One such is globalisation, another European integration. In addition sectoral trends are seen as important, for instance an emphasis on quality over quantity, the growing power of shipping companies, the increase in container cargo and the development of a hub-and-spoke system in European air traffic.

The second element in the 'mainport' discourse is the role of the state in this process. Although the discourse places great emphasis on market forces, the role of the state is quite significant. The state should provide the conditions for the development of mainports, not only in terms of regulations, but also through the provision of investment in infrastructure. This expenditure is seen as strategic, long-term investment, which does not necessarily have to show a return (Boom and Metze, 1997).

The third element of the 'mainport' discourse is the emphasis on time, the speed of decision making and of starting construction. 'No time to waste' is the motto, based on the idea that the international economy is unstable, that other countries are developing similar strategies and that the Netherlands is lagging behind. A report from the transport lobby (*Nederland Distributieland* - NDL) stated firmly: 'We need extra financial resources and faster decision making, since the Netherlands are lagging behind the neighbouring countries. Our position as a transport nation is threatened if investments are not made soon' (Nederland Distributieland, 1992). Over the last ten years the issue of speed of decision making regarding large-scale infrastructure has been the topic of parliamentary debate and study in the Netherlands. In the next section the institutionalisation of the mainport discourse will be described.

4.3.3 The institutionalisation of the mainport discourse

Discourse institutionalisation concerns the way in which a discourse is translated into practice (regulations, working routines, procedures etc.). It implies that the introduction of a policy concept has to be followed by some kind of institutional change in order to achieve a different policy outcome. The mainport discourse was quickly institutionalised, in three different ways: the introduction of legislation to speed up decision making, the new role of the ICES and the establishment of the Fund for the Strengthening of the Economic Structure (*Fonds versterking Economische Structuur* - FES).

In 1994, the 'Nimby Act' (*Nimbywet*) and the 'Track Act' (*Tracéwet*) (on planning procedures for new transport infrastructure routes) came into force. These regulations, with reference to 'national interests' on the one hand and 'not-in-my-back-yard' objections on the other, were intended to accelerate normal procedures which ensure public participation, legal protection and consultation. As stated in the Tracéwet a 'track decision' for projects of 'national importance' (*Tracé-besluit*) has far-reaching impact on future decision making, because it is binding on lower tiers of government (provincial and municipal councils). The act is therefore in conflict with the traditional method of decentralised spatial planning in the Netherlands. The 'track decision' is an explicit spatial plan which does not leave much room for incorporating landscaping or noise reduction preferences of local actors. Furthermore, it gives central government the right to force local councils to co-operate in cases where no agreement on proposed new infrastructure projects can be reached. The opportunities for participation have been reduced as well. In the past several attempts had been made to introduce a similar procedure, but they all failed. At the beginning of the 1990s this latest attempt succeeded, as a result of the newly-emerged mainport discourse. In 1997 parliament debated the introduction of a special law on large infrastructure projects (*Handelingen Tweede Kamer 1996-1997 24690 nr. 4; TK 25000 nr. 28 XI*); the policy debate on speeding up decision making continued in 1998. Most prominent was the attempt to speed up the construction of a new runway at Schiphol airport, bypassing a number of existing procedures, but this failed. Furthermore the Track Act was re-evaluated by the Ministry of Transport and a number of changes were made.

The second way in which the mainport discourse was institutionalised was through the new role of the Interdepartmental Committee for Economic Structure (*Interdepartementele Commissie Economisch Structuur* - ICES). Traditionally, the National Planning Committee (Rijksplanologische Commissie - RPC) had functioned as the interdepartmental body which integrated various sectoral policies with spatial planning. With the emergence of the mainport discourse the

ICES became more prominent as a body of policy initiation and co-ordination. The ICES consists of senior civil servants from the Ministries of Finance, General Affairs, Economic Affairs, Transport and Housing, Planning and Environment and is chaired and administered by the Ministry of Economic Affairs. It advises the government on the strengthening of the economic structure. In 1998 the new role of the ICES led to an interdepartmental conflict, which was resolved by giving ICES and RPC a joint office. The RPC was assigned the planning authority, ICES the distribution function.

Thirdly, the mainport discourse was institutionalised through the establishment of a number of funds. In 1991 it was decided that extra revenu from gas exploration would be reserved for strengthening infrastructure. This so-called *aardgasbaten-fonds* was replaced at the end of 1995 by the new Fund for the strengthening of the Economic Structure (FES). ICES deals with the distribution of the FES resources. The FES augments traditional forms of infrastructure, using formulated criteria for allocation. The distribution of resources follows a fixed formula: 85 per cent for infrastructure, 10 per cent for nature conservation and 5 per cent for 'other purposes'. Although the FES Act leaves open the possibility of investing in different forms of infrastructure (such as knowledge infrastructure, to promote services or clean technology), to date resources have predominantly flown to traditional forms. From 1994 onwards the Ministry of Transport has itself also administered a new fund: the Infrafund. This fund was established in order to be 'better able to carry the primary responsibility for the national infrastructure'. Expenditure for railways, motorways, canals, seaport infrastructure and airports is paid out of this fund. It derives its income from the general resources and the FES. The total budget for transport infrastructure has been raised significantly annually since 1990.

4.3.4 First experience: The Betuweroute

The first and most conspicuous project illustrating the growing importance of the mainport discourse was the planning of the Betuweroute, a new rail freight connection between the port of Rotterdam and Germany (See Figure 4.1).

The Betuweroute idea was favoured by both the port of Rotterdam and Dutch Railways (Nederlandse Spoorwegen, or NS) and it was placed on the political agenda in 1990. It was to be the first large engineering project since the construction of the Oosterscheldedam and the first major rail project in more than a century. The main rail network as it exists today has been in place since 1875 (Jonckers Nieboer, 1938).

Figure 4.2: The planned Betuweroute in the Netherlands

The initial decision to construct a new rail freight connection was a typical example of top-down policy making. It provoked protest from an unusually large number of local authorities and residents, who saw the project as a threat to their local environment. These protests were not just against the plans themselves, but also to a significant degree against the way they were presented by NS. Detailed plans were prepared by the Ministry of Transport and NS and presented to the local authorities in May 1991. Unfortunately, the plans were based on outdated maps on which several residential areas were missing. The autocratic style of the Minister of Transport Hanja Maij-Weggen, also provoked a great deal of resistance. During a public hearing one local resident said: 'You seem so down-to-earth and detached; what will our children say later on?' The minister replied: 'It is not my job to consider that sort of question' (Boom and Metze, 1997: 11). Most people felt that the formal hearings were meaningless, because the decisions had already been made. In fact, many officials who attended the public hearings were also dissatisfied. The issue of the Betuweroute has been particularly prominent over the years, especially at moments of decision making and during election campaigns. From the start local authorities formed coalitions and took on the role of protester, a new development in Dutch administrative culture. From time to time they used their position to provide local activists and the media with strategic information. They also mobilised alternative expertise and co-ordinated action with residents groups. Members of parliament felt that drastic changes were needed in the design of the route in order to safeguard local living conditions and protect wildlife.

The need for and purpose of the project itself were discussed exhaustively, but only when the key actors had already committed themselves to the project.

As one official put it: 'Costs and benefit calculations were never made for roads either and all of a sudden the government asked for an assessment of the need for and purpose of the project'. Over the years a number of cost-benefit studies were carried out, based on varying assumptions about the increase in the volume of freight and the shift in 'modal split'. These studies reached very different conclusions. Actors who opposed the Betuweroute tried to widen 'the discursive space' of the debate by arguing that an improvement in inland shipping would be a better way of solving transport problems, while others argued that in an age of information technology, transport and distribution were outdated methods of ensuring economic development. In May 1995, after the new government had appointed a committee to review the decision taken in 1993 by the previous government, it finally swept all studies aside and argued that the decision was simply a strategic one. The strategic advantage of the port of Rotterdam was at stake and this was more important than the specific outcomes of any cost-benefit studies. This argument, combined with the late stage at which fundamental questions about the need for and purpose of the project were raised, led many to believe that the quality of decision making had been inadequate. The Administrative Court (*Raad van State*) which decides on controversies regarding public decisions, received a record number of complaints.

The final decision to construct the line, at a cost of 8.3 billion guilders, much higher than the original estimates, was taken in November 1996. In 1998 public debate started again, when a number of academics called for a postponement of the construction and a rethink on the decision because new facts had come up. Parliamentary opposition parties agreed with this, but the government and the ruling parties thought otherwise.

4.3.5 Going the other way? Experiments in interactive planning

In sections 4.3.2 and 4.3.3 the emergence of a new policy discourse was described. The mainport discourse was the institutional choice to accommodate specific social and political demands. Its strong emphasis on economic goals together with the emphasis on time pressure led to a preference for central planning and the speeding up of procedures. In other words, the discourse called for a specific policy arrangement. It seems a perfectly straightforward case of policy renewal: a new discourse leads to a new policy coalition as well as to new policy practices (see also Hajer, 1995).

This is, however, only half the story of infrastructure policies in the 1990s. A completely different development occurred simultaneously: a large number of policy initiatives and experiments in interactive planning were initiated. The negative experience of the Betuweroute was an important incentive to change in this respect. Decision-making processes had reached deadlock and did not have sufficient support from the public and lower tiers of government. Furthermore, the Ministry of Transport felt that it was no longer able to deal with all problems, such as safety and traffic congestion, for which it was held to be responsible. Until the beginning of the 1990s the Ministry considered that it had all the answers to questions regarding water engineering, traffic, mobility and infrastructure. Society played mainly a signalling role: societal wishes were to be translated by the state into the technical design of infrastructure or to be integrated in transport policies. At the beginning of the 1990s the awareness grew within the Ministry that it could not on its own solve all the problems it encountered, that the Ministry itself might be part of the problem, and that conventional forms of policy implementation were no longer sufficient.

As mentioned before, both officials and citizens saw public participation in the late 1980s as unsatisfactory. In the words of Veldboer: 'it did not stand the test of time. It was a form of public management that fitted the culture of the 1970s, but not the commonsensical, individualised culture of the 1990s. Citizens wanted to be taken seriously from the beginning of a policy process and did not want to be invited into a smoky backstage conference room at the end of the process. It appeared that there is no need for ritual dances. The strongly confined forms of participation at the end of policy processes brought about prejudice and stiffening rather than support and creative contributions' (Veldboer, 1996: 15-16). This prejudice and stiffening in turn evoked an image of participating citizens as troublesome, egocentric, conservative and shortsighted. To put it briefly, there was a strong need to explore new ways of policy making. Recent initiatives can be found in five areas.

First, in 1993 the government asked the Dutch Scientific Council for Government Policy (*Wetenschappelijke Raad voor het Regeringsbeleid* - WRR) to investigate how existing decision-making problems in the infrastructure field could be overcome. The WRR report had considerable impact. Although the government did not issue a separate bill for decisions on large-scale infrastructure as the council had recommended (WRR, 1994), it did introduce a new consultation method: the 'need and purpose debate' (*nut en noodzaak discussie*). So far this method has been applied twice, in debates on extension to the port of Rotterdam and to Schiphol airport.

Secondly, the way formal public hearings were organised was changed. The Consultative Committee on Roads (COW) was relieved of its duty to organise hearings and this task is now performed by a subsection within the Ministry.[1]

---

[1] The COW was abolished after changes were made to the system of advisory councils in 1992, when advisory and consultation functions were separated. Consultation now takes place in the

Thirdly, in 1993 Rijkswaterstaat established InfraPlan (initially called InfraLab), an innovative forum in which, in interaction with citizens, new methods and new ideas could be developed. The purpose of InfraPlan was twofold: to generate a new approach to planning and decision making and to develop a long-term perspective in which a number of options for mobility and infrastructure could be explored (Van Enthoven, 1996).

Fourthly, a new practice was established whereby the development of transport infrastructure would be achieved in co-operation with lower tiers of government through the use of Administrative Committees (BBGs). These committees oversee the initital phase of the planning process in which the advantages and disadvantages of several alternatives are studied.

Finally, and perhaps the most important development, is that in recent years policy initiatives have no longer been confined to the Ministry of Transport. Local governments have empowered themselves and have pursued initiatives of several kinds. Some try to integrate the planning of different infrastructure projects, whereby redesign of the area is promoted rather than the landscaping of individual projects (Bestuurlijke begeleidingsgroep IVIL, 1997; De Waard et al, 1998). In another case, in the municipality of Renkum near Arnhem, a local councillor took the initiative in the track design for the planned high-speed train to Germany. Renkum organised expertise, local support and financial resources to implement its own ideas. In other words, although the Ministry of Transport remains the most important actor in the field, it is no longer the sole policy maker. Initiatives now come from various sides.

In the 1990s we see on the one hand a renewal of policy content (the emergence of the mainport discourse) and derived from that a strong emphasis on speeding up and centralising decision making, and on the other hand a renewal of public participation. This renewal does not stem from specific policy preferences but is a result of policy experiences in the field: the Ministry of Transport's inadequate ability to solve problems, felt most deeply in the experiences with the Betuweroute. The outcome of this is the co-existence of different policy arrangements which interact smoothly with each other. The empirical question therefore is how does the renewal of public participation work under the conditions of the institutionalised mainport discourse and the existing working practices? This question will be addressed in section 4.4.

Consultative Body for Transport Infrastructure (*Overlegorgaan Verkeersinfrastructuur* - OVI) and the advisory function is carried out by the Traffic and Watermanagement Council (*Raad voor Verkeer en Waterstaat*).

4.4 NEW POLICY ARRANGEMENTS IN PRACTICE

In the 1990s some policy actors have a strong wish to speed up decision making for transport infrastructure projects, but at the same time an authoritarian style of public policy making is no longer an effective means of reaching policy goals. In other words, the institutional preference of the mainport discourse does not seem to fit into the contemporary institutional and societal context. Looked at from the opposite angle, the urge to renew public participation is logical, given the implementation problems which have occurred and the lack of legitimacy of decision making. However, the communicative character of this renewal contradicts both the preferred content and the organisation of the policy field as expressed in the mainport discourse and these two tendencies therefore have opposite effects. In this section these contradictory effects will be examined by looking at three recent policy initiatives: the northeastern rail link to the Betuweroute, the 'need and purpose debate' on extending the port of Rotterdam, and InfraPlan projects.

4.4.1 Northeastern Link to the Betuweroute (NOV)

As a follow-up to the Betuweroute an extension of the rail connection to Germany, the northeastern link (*Noordoostelijke verbinding* - NOV), was planned, to accommodate both international and domestic freight (see Figure 1). As a result of earlier experiences, a new planning procedure was followed. An informal steering committee, comprising all public actors at the local, regional and national levels, was set up before formal procedures started. From July 1996 to May 1997 a study was conducted, exploring ten alternatives on the basis of a problem definition agreed by all public actors, which led to the elimination of five alternatives. The remaining five were then subjected to an environmental impact assessment and an accompanying feasibility study.

The internal co-ordination of the project at the Ministry of Transport was quite different as well. Instead of Dutch Railways and the National Transport Department in The Hague, a regional department of the Ministry did the planning of the project. It organised regular talks during the planning process. Private individuals and environmental groups were consulted (to their satisfaction) and every six months bilateral talks were held with the local councils and property administrators. Communication expertise was brought in to ensure better contact with residents.

There was, of course, quite a difference between the position of the members of the administrative steering committee and the other actors. Because an informal administrative consensus had been formed, the possibilities for societal actors and local residents to influence the final outcome were reduced. The minister gave priority to the consensus at the administrative level over the support of the private actors in the region. The reason for this was that the construction of the rail link had already been approved by the parliament as part of a package deal with the construction of the Betuweroute, a treaty with Germany on this subject had been signed and promises made to the provincial council of Gelderland. This led to a situation where the Ministry had to carry through the project as an administrative co-production, but without far-reaching forms of public participation. In 1999 the project was postponed indefinitely due to lack of financial resources and public support.

4.4.2 Exploratory Phase re Space Constraints in Mainport Rotterdam (VERM)

The *Verkenningsfase Ruimteprobleem Mainport Rotterdam* (VERM) was conducted from April to December 1996. The project was the first to be organised in accordance with the new procedure for large-scale infrastructure projects mentioned earlier. The essence of this procedure is that a 'need and purpose debate' precedes the formal planning procedure. The starting point is a 'commencement decision' document from the transport minister. In the case of the VERM process this stated that no particular solution should be promoted, but answers should be given to the following questions:
(1) What is the nature, size and urgency of the shortage of space?
(2) What can be achieved for the economy and the local environment by solving the shortage of space?
(3) Which possible solutions should be considered?

The document declared that it was the government's objective both to strengthen the position of the port of Rotterdam and to improve the local environment. The project consisted of four phases: consultation, problem analysis, possible solutions and decision making. Various actors played a role in the process - citizens, local officials, interest groups and experts. Public participation took the form of written and telephone surveys, round table conferences (on the conditions for possible harbour extensions) and regional and national workshops (on benefits, needs and possible solutions). A number of initial talks were held with interest groups to explore the problem.

Many of the participating citizens were satisfied about the round table conferences, although interest groups were somewhat less so (VERM, 1997). In December 1996 the results of the discussions were published. At the end of the VERM process the environmental movement expressed dissatisfaction and argued that they had only been allowed to discuss a limited number of issues and that existing policies had formed the starting point for discussion. Moreover, they criticised the Ministry of Transport by stating that the underlying data for the discussion originated from the Rotterdam Port Authority, an actor with a clear interest in a specific outcome.

Another point of criticism was that it took a very long time before the government was willing to ask the Commission for Environmental Impact Assessment for advice on the exploratory phase. The Commission then concluded that 'relevant questions from the environmental movement were not adequately considered'. Furthermore, it remained unclear what was meant by 'improvement of the local environment'. Twice a mediator had to be appointed to manage conflicts. The results of the mediator were confidential. According to the environmental movement this damaged the open character of the planning process.

During the parliamentary debate on the annual budget the transport minister made the following remarkable statement: '*in the need and purpose debate we discuss the shortage of space in the Port of Rotterdam and nothing else. Sometimes I have the feeling it is possible to conclude that there is no such shortage, but that is not at stake here*' (*Handelingen Tweede Kamer, 1996-1997 25000 nr. 4*). After the Central Planning Bureau (CPB), the country's economic planning office, had questioned the Minister's presumption of shortage of space, a broader approach was taken. This discussion emphasised the role of research institutes, since the size of the shortage was the key issue in the processes of complex decision making. The final decision made by the government was largely based on the CPB calculations: extending the harbour activities, but on a smaller scale then planned, and with the provision of extra areas for nature in compensation.

4.4.3 InfraPlan

In 1993 the Ministry of Transport established InfraPlan as an innovative forum in which, in interaction with citizens, new methods and new ideas could be developed. The forum was established to tackle the problems of traffic congestion, but the support for and subsequent, wide application of the method was partly a result of the negative experiences with the Betuweroute. It took too long to carry out projects; new solutions had to be considered and existing problems needed to be redefined.

The InfraPlan method is guided by three central principles: consultation of the end-user, co-operation between government and citizens throughout the process and a time limit of one year. The method consists of three phases, namely problem definition, developing possible solutions and designing plans for the chosen solution. The administrative authority keeps final responsibility and makes the decisions at the end of each phase. During the first phase actors have to communicate with all other parties, in order to develop common problem definitions. The results form the input for a written or telephone survey. This method establishes direct contact with the end-users of infrastructure, bypassing traditional consultation procedures with interest groups. For instance, on traffic congestion issues contact is made with individual car drivers and local residents rather than with motoring organisations.

Interactive planning methods for the construction of roads have been used in several provinces. Reactions have been positive, especially from participating citizens (Van Enthoven, 1996: 393). A negative point is that citizens have to wait between a year and eighteen months before any action can be taken, due to the fact that traditional procedures still apply as well. Within the administration opinion is divided. The new way of working requires that specialised and skilled professionals adopt a new attitude, postpone making their own judgements, and be open to new solutions. There are considerable differences in the enthusiasm for and faith in the new approach shown by the regional departments.

4.4.4 *Evaluation*

An evaluation of these three policy initiatives shows that, in practice, they are less open and more regulated than seemed to be the case at the start of the projects. In VERM, existing policies set limits on the discussion, and tension was felt between the informal process and the formal political sphere. The VERM project team adopted a more open attitude and was willing to engage in serious discussion, but senior civil servants were not (Andersson, Elffers and Felix, 1997). Moreover, commitment to the process from government and parliament was limited. This experience suggests that new patterns of policy making can only be used if there is sufficient support from the political sphere. Another problem which arose was the accusation by the environmental movement that specific interests guided the selection of data on which the discussion was based. One of the difficulties here seems to be the role of information and the acquiring of knowledge during the debate. Knowledge is usually seen as a source for answers, rather than as part of the decision-making process. At the end of the process the environmental movement argued for a joint effort to establish a research programme based on collective problem definitions. During the second half of 1997 a second 'need and purpose debate', on the future of Dutch air transport, was held. In this case information played a much more prominent role, because a scientific discourse was explicitly linked to the public debate.

InfraPlan projects have engendered only moderate enthusiasm, due to the standard bureaucratic way of working. In this case the problem seems to be that it is not easy for officials to accept the pluralism of society and the different views of reality held by various actors. It is therefore difficult for these professionals to benefit from interaction with the various actors and not to see them as rivals. Furthermore, existing formal procedures, which follow the informal InfraPlan process, prevent actors from benefiting from the momentum created by InfraPlan. Previous commitments and the emphasis on administrative consensus limited the effectiveness of the NOV process. A partial response to this kind of institutional obstacle is flexibility and contingency planning in long-term spatial and transport plans. The extent to which this is possible is beyond the scope of this chapter.

Not only do we see policy initiatives to be less open then expected, from the perspective of the institutionalised mainport disourse it is also questionable whether regulations to speed up decision making will be as effective as hoped. Local authorities can operate strategically within the legal framework. In fact, the Track Act is a trigger for local government to do so (Aarts, Huigen and Van der Rijdt, 1995). Due to the predominance of decentralised spatial policy making, the national government also depends on local governments for all sorts of practical but vital information and services (issuing licenses, providing information on soil conditions, etc.). At the same time, the atmosphere that it is hoped will be created by the new forms of participation — trust, openness, fairness — is damaged by the use (or threat of use) of the Track Act. Equally, the state gives ambiguous signals to its citizens. While their participation in a number of experiments is encouraged, citizens are at the same time accused of making 'nimby' objections and the opportunities for voicing their opinions are reduced.

In the individual projects we can see how to a certain extent the diverse logic of each policy arrangement outplays the other. This is not to say that by definition the two forms of policy making cannot be successfully joined. It appears possible to create policy arrangements for transport and infrastructure in which serious participative elements are combined with the ability to take decisions when it is felt necessary. However, the prime precondition for this is a common understanding within the Ministry of Transport between 'interactive planners' on the one hand and 'centralists' on the other. On the basis of such a common understanding an attempt could then be made to integrate the two types of policy arrangement.

4.5 CONCLUSION

Compared to the policy arrangements that were dominant in the 1950s and 1960s, infrastructure policies nowadays are opposed more often and more actors are engaged in the debate. Compared to the 1970s, the Ministry of Transport has become much more aware of the negative side effects of mobility and policies and mechanisms are designed to cope with these negative effects. The coalition, which promotes infrastructure, has enlarged as well, compared to the 1980s. On the one hand a large number of procedures are designed to engage in open discussion with the public, on the other hand there are clear targets to be met and procedures are intended to meet these targets.

From the institutionalist perspective described in the first chapter of this book one can argue that the renewal of public participation is a sign of reflexive modernisation. A dominant state actor is no longer able to handle its own problems and is forced to call for institutional change (Pestman and Van Tatenhove, 1998). New arrangements in Dutch infrastructure policies clearly have some characteristics of reflexive policy making, in which in a newly-created institutional setting policy problems and solutions are defined by means of co-production. The institutional preference of the mainport discourse on the other hand seems to stem from a different form of political modernisation. The rule-directed politics advocated by the mainport discourse and actualised in the Track and Nimby Acts is an attempt to centralise decision making.

Overall, however, the current situation resembles the picture outlined by Beck: a co-existence of rule-directed and rule-altering politics. The limited openness of the new initiatives can only partially be explained by reference to problems of transition. While the rise of the negotiating state as argued by Beck can be acknowledged, it is possible to take the argument one step further by looking at the situation as it is. As argued in chapters 2 and 3, what is characteristic of the present situation is not the transition of policy arrangements from uniformity and control to consensus building and negotiating, but the existence side by side of different arrangements. Politics is both hierarchic and horizontal, and governability is both uni-centric and multi-centric. While Beck sees this as a temporary phase of transformation, it can be argued that this ambiguity is inherent in the present state of policy making. The challenge then is to integrate different arrangements from different conceptual backgrounds. For instance, in a report from the Scientific Council for Government Policy, a proposal was made to take, after public debate, decisions on large-scale infrastructure at national level. The design of such infrastructure should be at regional level and in this secondary process regional public and private actors should be involved (WRR, 1998). National government would have to make a fixed budget available and provinces and municipalities would then be entitled to draw up a binding land use plan. This is just one possible example of how it might be possible to combine a plurality of policy arrangements in order to overcome contradictory tendencies.

REFERENCES

Aarts, J.H.W.M., J. Huigen and D.H. van der Rijdt (1995), Snelheid van uitvoering onder Tracewet: op weg naar pervertering van de verhouding tussen rijksoverheid en medeoverheden?, *Bestuurswetenschappen*, 49(4), pp. 278-292.

Andersson Elffers and Felix (1997), *Ruimte voor discussie - evaluatie van de nut- en noodzaakdiscussie Ruimtetekort Mainport Rotterdam*, Utrecht: AEF.

Bestuurlijke begeleidingsgroep IVIL (1997), *IVIL: Integrale Verkenning Infrastructuur 'De Liemers'*, Arnhem: Bestuurlijke Begeleidingsgroep IVIL.

Boom, H. and M. Metze (1997), *Slag om de Betuweroute: het spel langs de lijn*, Amsterdam: Balans.

De Heer, J.M. (1991), *Strategie en verandering in organisaties onder druk: een onderzoek naar het besturen van herpositionering en transitieprocessen in organisaties van Verkeer en Waterstaat*, Den Haag: VUGA.

De Waard, R., C.A.M. Luttikhuis and J. Wabeke (1998), Het kleine denken in de grote infrastructuurprojecten, *Stedebouw en Ruimtelijke Ordening*, 79(4), pp. 13-16.

Hajer, M.A. (1995), *The Politics of Environmental Discourse: Ecological Modernization and the Policy Process*, Oxford: Oxford University Press.

Huberts, L.W.J.C. (1988), *De politieke invloed van protest en pressie: besluitvormingsprocessen over rijkswegen*, Leiden: DSWO Press.

Jonckers Nieboer, J.H. (1938), *Geschiedenis der Nederlandsche Spoorwegen 1832-1938*, Rotterdam: Nijgh and van Ditmar.

Knoester, A. (1989), *Economische politiek in Nederland*, Leiden: Stenfert Kroese.

Ministerie van Economische Zaken (1976), *Economische structuurnota*, Den Haag: SDU.

Ministerie van Economische Zaken (1990), *Economie met open grenzen*, s-Gravenhage: SDU.

Ministerie van Economische Zaken (1995), *Ruimte voor regio's: het ruimtelijk-economisch beleid tot 2000*, Den Haag: SDU.

Ministerie van Economische Zaken (1997), *Ruimte voor economische dynamiek: een verkennende analyse van ruimtelijk-economische ontwikkelingen tot 2020*, Den Haag: Sdu.

Ministerie van Verkeer & Waterstaat (1981), *Structuurschema Verkeer en Vervoer, Deel E*, Den Haag: SDU.

Ministerie van Verkeer & Waterstaat (1988), *Tweede Structuurschema Verkeer en Vervoer Deel A*, Den Haag: SDU.

Ministerie van Verkeer & Waterstaat (1990), *Tweede Structuurschema Verkeer en Vervoer, Deel D*, Den Haag: SDU.

Ministerie van Verkeer & Waterstaat (1992), *PKB Betuweroute deel 1*, Den Haag: Ministerie van Verkeer & Waterstaat.

Ministerie van Volkshuisvesting en Ruimtelijke Ordening (1973), *Oriënteringsnota*, Den Haag: SDU.

Ministerie van Volkshuisvesting, Ruimtelijke Ordening en Milieubeheer (1985), *Ruimtelijke Verkenning Hoofd Infrastructuur*, Den Haag: RPD

Ministerie van Volkhuisvesting, Ruimtelijke Ordening en Milieubeheer (1988), *Vierde Nota Ruimtelijke Ordening deel A*, Den Haag: SDU.

Ministerie van Volkshuisvesting, Ruimtelijke Ordening en Milieubeheer (1990), *Vierde Nota Ruimtelijke Ordening Extra, deel 1*, Den Haag: SDU.

Ministerie van Volkshuisvesting, Ruimtelijke Ordening en Milieubeheer (1990), *Vierde Nota Ruimtelijke Ordening, deel D*, Den Haag: SDU.

Nederland Distributieland (1992), *Concurreren met infrastructuur: inhaalplan voor de Nederlandse infrastructuur: baanbreken naar Europa*, Den Haag: NDL.

Pestman, P. and J. van Tatenhove (1998), Reflexieve beleidsvoering voor milieu, ruimtelijke ordening en infrastructuur: Nieuwe initiatieven nader beschouwd, *Beleidswetenschap*, 12(3), pp. 254-272.

Raad van Advies voor de Ruimtelijke Ordening (1970a), *Advies inzake het betrekken van de bevolking bij de vorming van het ruimtelijk beleid, uitgebracht aan de minister van Volkshuisvesting en Ruimtelijke Ordening*, Den Haag: Staatsuitgeverij.

Raad van Advies voor de Ruimtelijke Ordening (1970b), *Advies over de openbaarheid bij de voorbereiding van het regeringsbeleid inzake de ruimtelijke ordening*, Den Haag: SDU.

Van Delden, A.T. (1985), Adviesorganen, R.B. Andeweg, A. Hoogerwerf and J.J.A. Thomassen, eds., *Politiek in Nederland*, Alphen/Rijn: Samson H.D. Tjeenk Willink, pp.148-177.

Van Enthoven, G.M.W. (1996), Infralab, impuls voor open planvorming en creativiteit, *Bestuurskunde*, 5(8), pp. 392-399.

Van der Cammen, H. and L.A. de Klerk (1993), *Ruimtelijke ordening: van plannen komen plannen, de ontwikkelingsgang van de ruimtelijke ordening in Nederland*, Utrecht: Het Spectrum.

Veldboer, L. (1996), *De inspraak voorbij: ervaringen van burgers en lokale bestuurders met nieuwe vormen van overleg*, Amsterdam: IPP.

VERM (1997), *Eindverslag werkateliers (VERM)*, Den Haag: Projectgroep VERM

VERM (1997), *Verkenningsfase Ruimtetekort Mainport Rotterdam*, Den Haag: Projectgroep VERM.

Wetenschappelijke Raad voor het Regeringsbeleid (1994), *Besluiten over grote projecten*, Den Haag: SDU.

Wetenschappelijke Raad voor het Regeringsbeleid (1998), *Ruimtelijke ontwikkelingspolitiek*, Den Haag: SDU.

5. Nature Conservation Policy: Transboundary Arrangements

DIANA DE JONG

5.1 INTRODUCTION

Nature conservation policy in Europe has clearly changed and developed in the last few decades. Policy arrangements evolve both on the European level and within the member states of the European Union. The Fifth Environmental Action Programme (1992, the most recent) incorporates insights from new scientific, social and political discourses. A clear expression of this process can be found in the discourse on 'ecological networks'. The central policy goal in this discourse is the realisation of a coherent European network of protected areas, entitled 'Natura 2000'. Developed from scientific concepts such as island theory and metapopulation dynamics, ecological networks aim to provide the physical conditions that are necessary for species populations to survive in a landscape that to a greater or lesser extent is also exploited for economic activities. In post-modern discourse on the decline of nature, the reduction in area of many natural and semi-natural habitats and the fragmentation of large habitats into small 'islands' that become isolated from each other are held to be the main causes of this decline (Bennett and Wolters, 1996). The concept of ecological networks applies to the European and the national as well as the regional level. Political actors at different levels develop policy measures with the intention of connecting scattered nature areas.

Awareness of a bio-geographical, transboundary dimension in nature policy has begun to emerge in the political domain since the early 1980s. As a consequence of the recognition of bio-geographical rather than political borders, ecosystems are considered to be interconnected across state borders. In the background of the concept of ecosystems without borders lie the facts that the importance of national borders within the European Union is decreasing and that globalisation makes countries more interdependent. In the political discourse co-operation is more and more held to be necessary, not only because of ecological connections across borders but also for political and administrative reasons. Co-operation is becoming more important because the process of European integration modifies the economic and political position of border regions, with these regions evolving from remote, relatively underdeveloped and peripheral areas to central transition zones within the European Union. European initiatives such as the Interreg programme also stimulate this process. This policy intends at 'providing stimuli for the foundation and development of co-operative networks across internal borders, and to link these networks to larger community networks' (European Commission, 1990 in: Van Houtum, 1998).

Changing discourses on nature, the different European context in which border regions function and the need for transboundary co-operation as a consequence of changing power relations all demand new policy arrangements. Transboundary policy on nature conservation is developing in a small number of border regions within the EU. Co-operation with regard to nature conservation is not directly stimulated or initiated by European policies or directives, rather transboundary policy arrangements are formed as a result of national initiatives, especially by lower tiers of government such as provinces, municipalities and regions. Within these regional networks, however, actors have to act within the context of national and European arrangements and European policies will be discussed in this chapter in this light. Transboundary policy processes in the field of nature conservation are considered to be the result of interactions between interdependent actors, who in their interaction refer to policy arrangements at European and national levels. The central question in this chapter is how transboundary policy processes in the field of nature conservation work and how and to what extent they are influenced by European and national policy arrangements.

A German-Dutch nature conservation project, the *Gelderse Poort*, will be analysed in terms of a newly-developing policy arrangement. The formulation of a policy for this transboundary area requires co-ordination, consultation and co-operation between governmental actors on different levels and between political and private actors. Such policy processes often are complicated or even frustrated by the existence of divergent or conflicting discourses about the problems involved, the best solutions and the actors who should be involved. These aspects are reflected in the characterisation of policy arrangements in chapter 3. Following the theoretical assumptions of this book four aspects - policy coalitions, division of resources, rules and environmental discourses - distinguish policy arrangements. Considering the specific features of transboundary nature policy, particularly the fact that two countries have little obligation towards each other in this policy field, the emphasis in this chapter is on discourses and the development of coalitions. In section 5.2 some general remarks about transboundary policy arrangements are made. Section 5.3 outlines policy arrangements on both European and national level which have implications for the policy process in the *Gelderse Poort*. The context in which the *Gelderse Poort* arrangement is embedded then becomes clear. The influence of different definitions of the policy problem, the targets and the problem-solving strategies of the policy process will be analysed on the basis of a reconstruction of the policy coalitions and their discourses in section 5.4. In this chapter the development of discourses and patterns of relationships in the *Gelderse Poort* arrangement will be analysed from the beginning of the policy process in around 1990 until 1998. On the basis of this analysis the policy process and the outcome can be understood (section 5.5).

5.2 TRANSBOUNDARY POLICY ARRANGEMENTS

In recent years, the need for a transboundary approach to nature conservation has become more acknowledged and the transboundary dimension has become part of the discourse at different levels. This development was stimulated by the changing economic and political status of border regions. Due to their geographical location and political function, the economic development of these regions has often been hampered through the ages. In the past nation states generally promoted their core areas and kept the border areas as a kind of no-man's land. As a consequence, the classical characteristics of these border areas are distance, difference and dependence. A border area is located at some distance from the politically and economically dominant centre(s). A border area is, secondly, different from the central areas on one or more counts. While the degree of distinctiveness will vary, being to some extent a function of distance and dependence, there will invariably be some minimum level and sense of separate identity. Thirdly, a border area is dependent upon one or more centre(s), either politically or economically (Rokkan and Urwin, 1983). This, combined with low population density and a relatively underdeveloped regional economy, means border areas are often of high ecological value. Because of European integration these values are under pressure.

Over the last few decades the importance of state borders has seemed to diminish. Although one should distinguish between actual developments and the ideals and rhetoric of European integration, the economic integration of the European Union affects the importance of state borders (De Swaan, 1994). One of the goals and effects of the EU policy is the increase in international economic activity and the transboundary transport of goods etc. As a consequence border regions face more intensive traffic and trade and they seem to become attractive regions for new investment in infrastructure and in sectors such as stocking, transport and expedition. Unfortunately, the EU policy on economic integration is not accompanied by a simultaneous formulation of transboundary policies on the environment and nature. While it is true that the EU is elaborating its own environmental policy, transboundary environmental policies have yet to be developed. National borders still seem to be an important obstacle to the designing and implementation of transboundary policy on nature (De Jong and Leroy, 1996). Recent initiatives for transboundary policy arrangements have originated at the regional or national levels.

In these policy processes, and in policy processes in general, a government operates in a setting of interdependent actors and coalitions. Government is not the only, nor the central steering authority (see, among others, Van Vught, 1986; Van Twist en Termeer, 1991). The policy-making process is viewed as a process in which the different goals of several organisations have been intertwined to become an acceptable compromise. Separate actors may participate in the policy process for their own reasons (Klijn and Teisman, 1991). Especially in international or transboundary policy processes, these reasons often differ as a result of the institutionalisation of policies in different countries.

The growing international interdependency forces actors in different countries, willingly or unwillingly, into some form of co-operation. When an actor can obtain his or her preferred outcome by making independent decisions there is no reason for transboundary co-operation. Whether or not actors choose to co-operate does not depend principally on whether the problem is one of absolute or relative gain (since the one can almost always be translated into the other), but on the prior question of how actors define the problem and its solutions (Stein, 1990). Policy arrangements exist because of the mutual dependence of actors. In transboundary policy processes the main reason for this interdependence is that actors need resources they do not have themselves. Policy arrangements are a result of negotiation between actors who are involved because they have 'problem-solving abilities' which cannot be excluded (Klijn and Teisman, 1991).

In the analysis of policy arrangements in the domain of nature conservation we focus on the interactions between actors, the coalitions which are formed as a result and the dominant discourses in these coalitions. This focus is inspired by insights from the configuration theory. Transboundary nature policy is an innovative policy domain and is at present mainly concerned with the construction of common concepts and meaning (discourses) in the formation of coalitions at regional level. To analyse such a process the configuration approach offers useful insights, for it stresses the importance of interactive policy processes at the operational level (De Laat, 1983; Maas, 1988; Bolk, 1989; Voogt, 1990; In 't Veld et al., 1991; Termeer, 1993).

One of the factors of structural social change is considered to be closely related to the continuous struggle by individuals to provide adequate frameworks for clarification and interpretation of (new) confusing or discongruent information which they experience and see (cognitive structures). At first sight social structures seem to be a result of cognitive structures (expressed in discourses). Depending on the concepts advocated, one looks for closer relationships with actors with whom ideas can be exchanged and attuned in a subjectively adequate way (coalitions). Secondly, social structures seem to be the forerunners of cognitive structures. One develops cognitive structures depending on the actors with whom one exchanges and attunes ideas. The one would seem to result from the other and vice versa (Bolk, 1989).

In this chapter policy coalitions in the *Gelderse Poort* process are reconstructed through an analysis of primary empirical data. The empirical research consisted of a content analysis of policy documents, minutes of formal meetings and the similar papers. These data were complemented by interviews with representatives of the parties involved in the policy network (De Jong, 1999).

5.3 POLICY ARRANGEMENTS IN EUROPEAN, GERMAN AND DUTCH NATURE CONSERVATION

5.3.1 European Policy Arrangements

Current European policy on nature conservation essentially consists of two pieces of legislation: Council Directive 79/409/EEC on the protection of wild birds (Birds Directive) which was adopted in April 1979 and Council Directive 92/43/EEC on the conservation of natural habitats and wild flora and fauna (Habitats Directive) which was adopted in May 1992. At the centre of the policy is the wish to create a coherent ecological network of protected areas across the EU, Natura 2000. The EU Habitats Directive is by far the most important resource in the European Commission's attempts to influence national policy arrangements in order to establish an ecological network. The Habitat Directive requires that 'natural habitat types of Community interest' and 'habitats of animal and plant species of Community interest', shall be maintained or restored to a good state of conservation (Jongman, 1995; Bennett and Wolters, 1996). National actors are responsible for the selection of areas, and implementation of the directives takes place through national policies. Actors at national level have to build coalitions with regional and local actors to enforce the policy within their country. Behind the two directives, however, is a policy which aims for a certain degree of harmony in the approaches taken by national governments in developing and implementing an ecological network, but implementation is indirect, through national and regional governments. In the discourse at European level the term Natura 2000 symbolises the conservation of natural resources from the year 2000, with the plan being realised in full by 2004 (Felton, 1996). At this moment several countries within the EU, including Germany and the Netherlands, have compiled (provisional) lists of nature areas which must be part of Natura 2000. Selection of areas is a matter for the member states, in consultation with regional governmental actors, nature conservation organisations and other interest groups. This process is time-consuming and implementation of the plan by 2004 is held to be impracticable.

Nature conservation has not been a major priority in EU policy in the past. Policies for species protection and the establishment of national parks, nature reserves and other protected areas have developed mainly at regional or national levels, and as a result substantial differences between different parts of Europe remain (Bennett, 1991). Different policies and concepts of nature conservation are encountered in particular in the state border areas. A coherent ecological network needs to cross these borders, however, since nature after all takes no account of borders. Co-ordination of national initiatives is therefore needed. Given the sovereignty of states, and the rules and procedures which flow from this principle, it is not conceivable that states will agree to far-reaching transboundary EU policy initiatives in nature conservation. Nature policy arrangements at national level are considered to have more impact on transboundary co-operation on nature conservation than those at European level. The empirical research on the *Gelderse Poort* case supports this assumption. In the next two sections the policy arrangements in the Netherlands and in Germany will be discussed.

5.3.2 Dutch policy arrangements

In the last two decades, the nature policy domain in the Netherlands has been subject to notable change. Since the early 1990s a policy arrangement has developed around the concept of ecological networks. Public as well as private actors have been working on translating of the ecological network discourse into policy concepts. An important resource within the political domain is the Nature

Policy Plan (*Natuurbeleidsplan*, 1990), in which the outlines of the ecological network are given. The central assumption behind this policy is that the future of ecosystems can be best guaranteed by linking areas with high nature values to form a coherent and robust ecological network of large-scale nature areas: 'Primary Ecological Infrastructure' (*Ecologische Hoofdstructuur*). Parts of the planned network are beyond the national borders, in Germany and Belgium, and the Dutch therefore intend to have crossborder connection of their nature areas. An example of such a project is the *Gelderse Poort*.

In Dutch discourses on nature policy nature development is a successful concept. Over the last few decades claims of being able to create 'spontaneous nature' and even 'new nature' have gained ground, motivated by new scientific concepts and insights, and initially advocated by biologists, ecologists and representatives of nature conservation organisations. The concept of 'spontaneous and pristine nature', free from most if not all human intervention, received strong impetus through the increasing competition between the established nature conservation organisations and a newcomer in the national network, the World Wide Fund for Nature (WWF). This new discourse on nature conservation opens opportunities for land to be converted into either protected landscapes or 'new nature' (Baerselman and Vera, 1989). In the Netherlands the emphasis is on the possibilities for the creation of new nature. At the end of the 1980s, proponents of nature development succeeded in building a coalition with governmental actors. Several pilot projects were initiated in co-operation between public and private actors and in 1989 the Dutch government selected the *Gelderse Poort* as a nature development project.

Most projects for nature development are situated on Dutch rivers, in particular the Rhine and the Meuse. Because of the dynamic ecosystems around rivers, positive results and measurable effects can be expected in a relatively short time. A striking assumption in this discourse is that the river water is considered clean enough for the ecological recovery of the river areas. It is thought many plants and animals will return once they find a natural landscape along the rivers where they can re-establish their habitats (Stroming et al., 1993). The WWF commissioned a study, 'Living Rivers', to explore these ideas.

In Dutch nature policy tasks and resources are allocated in a rather top-down and sectoral way. Since 1982, nature conservation has fallen within the competence of the Ministry of Agriculture, Nature Management and Fisheries (Baldock et al., 1987). The provinces depend on the national government in several ways and have to carry out their tasks within the framework of national policy. In the realisation of the ecological network, provinces are responsible but depend on the state for financial resources and concrete plans or proposals from the provinces have to be agreed by the minister.

5.3.3 German policy arrangements

In Germany nature conservation is mainly the responsibility of the *Bundesländer* (regional states). Although the federal government can pass some framework legislation, the *Länder* are responsible for the specific regulations and the implementation of environmental protection laws and their enforcement (nature conservation is a so-called '*eigene Angelegenheit*'). Here we discuss nature policy in North Rhine-Westphalia, because this state is involved in the transboundary co-operation of the *Gelderse Poort*. Environmental policy in North Rhine-Westphalia covers a broad range of sectoral tasks, including among other things environmental protection, nature conservation, waste management and nuclear safety.

Policy for rural areas in North Rhine-Westphalia encompasses more than the Dutch policies do. The objectives of nature conservation include the protection of the entire natural environment. The term 'Umwelt' (environment) is given a very wide interpretation, and the functions of environmental protection include nature conservation, landscape protection and so forth. Until 1993, no specific policy arrangements for the German part of the *Gelderse Poort* were formed. German nature conservation policy for that area in general is aimed at the conservation of the cultural landscape (*bäuerliche Kulturlandschaft*).

At the end of the 1980s, after German reunification, the environmental discourse stabilised. Unification and economic problems hampered innovations in nature policy arrangements, and in 1990s a new discourse slowly began to gain ground in the nature policy domain. The sixteen *Länder* reached general agreement on the basic principles and concepts of nature conservation (the so-called '*Lübecker Grundsätze des Naturschutzes*'). The Lübecker principles must act as an overall policy document, bridging divergent policies in the *Länder* and stressing common arguments and bases for action. There is agreement that areas earmarked for nature conservation should be categorised as core areas, buffer areas, network areas or areas for nature development (Heidenreich, 1993). This agreement, however, has not yet resulted in new, more concrete discourses at *Länder* level.

Nature conservation and (spatial) landscape planning are closely related and integrated in the so-called *Landschaftsplanung* (landscape planning). Landscape planning is embodied in the Federal Nature Conservation Statute (*Bundesnaturschutzgesetz*) and in laws of the *Länder* (in the case of North Rhine-Westphalia the *Landschaftsgesetz*), and should be regarded as an original and autonomous resource (instrument) of overall planning for rural areas (Heidenreich, 1993). Strategic, general goals for nature conservation have been formulated in Natur 2000, which is a sectoral plan for North Rhine-Westphalia. National nature policy and its central targets, however, are set out in a spatial plan, *Landesentwicklungsplan Nordrhein-Westfalen* (LEP; state development plan), in which nature interests are weighed against other objectives. The central goal in the LEP is the creation of an ecological network (*Biotopverbund*) across the state. Changes in rural areas and agriculture require an adaptation of policy arrangements in this field. An important objective is the promotion of less intensive and more organic farming, so that the agricultural landscape is preserved. The objectives of nature conservation and the preservation of landscapes include the protection of the entire natural environment. The basic principle underlying nature conservation policy in Germany is that interference with nature should be minimised and that any interference which does occur should be balanced by beneficial measures (compensation) in another location (Schmidt-Eichstaedt, 1996).

Important legal resources in the execution of German nature conservation policy are vested at the *Kreis* (county) level. In general the *Kreis* performs those tasks which either are beyond the financial capability of the *Gemeinde* (municipalities) or cannot be properly performed in a local framework (Paterson and Southern, 1991). In the domain of nature policy the *Kreis* is responsible for the formulation of landscape plans.

5.4 THE GELDERSE POORT ARRANGEMENT

The *Gelderse Poort* project is an example of a policy arrangement in which political actors aim at a transboundary approach to nature conservation. The *Gelderse Poort*, or *Niederrheinische Pforte*, covers an area of 210 km^2, half in the Netherlands and half in Germany. The area is part of the Rhine system, where the Rhine leaves Germany and flows into the Netherlands, dividing into a number of branches. In general this part of the river valley is characterised by levees, the presence of scattered nature reserves, agricultural areas both inside and outside the dykes, many clay quarries and various forms of recreational activity (Cörvers and Slot, 1995).

The initial goal of the project was large-scale protection and development of nature within the whole *Gelderse Poort* area. A project organisation was formed to work out the plans, with as the central actor the (Dutch) province of Gelderland. Policy formulation is originated by regional and national authorities and implemented by regional and local authorities. The *Gelderse Poort* project is embedded in Dutch national policy arrangements (such as the Nature Policy

Plan, the Further Development Plan for the River Area and several spatial plans) and is an offshoot of the nature development discourse. The institutional position of the project in the Netherlands clearly differs from that in Germany. In the Netherlands the claims to be able to create 'spontaneous nature' have been institutionalised. Areas are selected and developed on the basis of 'ecological reference'. This reference defines which kind of natural scenery would exist in the Netherlands if mankind had never influenced the landscape (Keulaertz, 1997). As a result of economic and political developments in agriculture, converting less productive agricultural land into either protected landscapes or 'new nature' began to be discussed. Influenced by new scientific knowledge, experiences in nature management and examples of successful nature development projects, doubts were then cast on the predominant definitions of nature, and in the 1980s a period of reflection and discussion on the aims and methods of nature conservation began (Baerselman and Vera, 1989).

Recently ideas about nature development, similar to the way it is proposed in the Netherlands, have been discussed in Germany. Traditionally nature development (*Renaturierung*) was associated only with compensation and the restoration of damaged areas (e.g. due to quarrying). In the dominant discourse on the protection of rural areas as expressed in nature and spatial policy, the preference is for the promotion of land use compatible with the natural environment. The availability of previously cultivated land for nature conservation is stressed and maintaining extensive use of grassland with a high ecological value is promoted (Roth, 1994).

The exact consequences and implementation of the nature development discourse at local level were not completely clear at the start of the policy process of the *Gelderse Poort*, they had to be developed during the process. Actors from different governmental levels and from several private organisations in both countries needed to express and exchange their policy goals for the area. Different conceptualisations of the most important functions, goals and concepts for the area were produced during the policy process.

Transboundary policy for the *Gelderse Poort* has to be worked out through consultation and interaction between civil servants and political representatives from different policy domains. The Dutch province of Gelderland took the initiative for the creation of a transboundary structure for consultation and decision making. After consulting the German Ministry of Environment, Spatial Planning and Agriculture (*Ministerium für Umwelt, Raumordnung und Landwirtschaft*, MURL) in 1992, it became clear that policy formulation and implementation for the *Gelderse Poort* had to be carried out at regional level. The central actor on the German side is the *Kreis Kleve* (county of Cleves). From 1993 onwards formal institutions for consultation and co-operation were put in place. In May 1993 a German-Dutch steering committee and a transboundary project committee were formed. The Germans, following the Dutch example, set up a group to act as a sounding board.

During transboundary interactions actors were confronted with different discourses on nature conservation. An important expression of this difference in nature policy in the Netherlands and Germany can be found in the way nature conservation is institutionally embedded. In the Netherlands the *Gelderse Poort* project is included in several discourses, which are partially expressed in policy documents, at different levels. Its inclusion in Dutch national spatial policy (initially the Fourth Policy Document on Spatial Planning, *Vierde Nota Ruimtelijke Ordening*) gives the project a strong basis in contemporary national policy arrangements on nature conservation. As a result considerable financial resources were put at its disposal. The area is also included in policies at provincial level, through the Further Development Plan for the River Area (*Nadere Uitwerking Rivieren Gebied*, NURG) and the spatial plan (*Streekplan*).

In Germany the institutional situation is different, as there is no binding federal spatial plan. The *Landesentwicklungspläne* (state development plans) function as instruments for the co-ordination of all policies with a spatial impact in each state (Schmidt-Eichstaedt, 1996). This policy domain does not initiate new discourses by proposing central concepts for a specific policy sector. Also the *Gelderse Poort* is mainly a local and municipal issue and is not embedded in national or provincial policy. Few policy arrangements specifically pay attention to the *Gelderse Poort* area. So the starting point on the two sides of the border is rather different.

In terms of coalitions in the policy domain we can identify three dominant coalitions, composed of government actors. Smaller coalitions, with less impact on the final outcome of the policy process, will not be discussed here.

German 'Kulturnatur' coalition

The German actors have formed a quite stable, coherent coalition during the whole policy process. Central actor and project leader is the *Kreis Kleve*. The federal Ministry of Environment, Spatial Planning and Agriculture, the *Regierungspräsident Düsseldorf*[1] (administrative district of Düsseldorf) and the

[1] The Bezirksregierungen, also called Regierungspräsidenten, are middle-level administrative authorities for the administrative districts (five in North Rhine-Westphalia) within the Länder. Each has a Regierungspräsident (district chief executive) appointed by the Ministerpräsident of the Land. The departments of the Bezirksregierung co-ordinate and supervise the implementation of federal and regional state legislation and regulations for the district (Schmidt-Eichstaedt, 1996).

municipalities concerned (the city of Cleves and the town of Kranenburg) for the most part share the vision of the *Kreis*. Discourses of the *Kreis* have given and still do give the lead to the other actors in deciding their position; they recognise the *Kreis* as the central and responsible decision-making actor. The *Kreis* has at its disposal legal instruments, mainly in the domain of landscape planning, for the realisation of nature conservation policy, and the actors consider the *Kreis* to have the most relevant resources to reach a transboundary agreement. The actors in this coalition are part of the German steering committee and represent Germany on the transboundary steering committee.

Typical of the discourse in this coalition is the wish to maintain historic cultural landscapes (*historischen, bäuerlichen Kulturlandschaften*) in which visual aspects of the landscape and the combination of agriculture and nature play a central part. The cultural landscape is considered valuable because of the way it looks and the species it contains. The agricultural landscape as it was at the beginning of this century is the most important reference point for this idea of nature. The animal species which prefer this kind of habitat are, among others, water birds such as geese and according to the central discourse in this configuration these species must be protected.

From the beginning of the 1980s this protection has been mainly organised through the implementation of an international convention. The German part of the *Gelderse Poort* is designated a wetland of international importance (*Feuchtgebiet internationaler Bedeutung*) under the Ramsar convention, an international convention on the conservation of wetlands. Because of the area's international status the Germans expected the Dutch to agree to this policy. However, the species such as geese which are valued in the German coalition do not have a place in the kind of nature the Dutch actors prefer, as will be explained below.

In the German policy the protection of indigenous species and their habitats has the highest priority, and agriculture is indispensable in the realisation of the kind of nature areas the Germans prefer. Therefore the policies strive for an interweaving of agriculture and nature. For the *Gelderse Poort* area this means a combination of agriculture and nature protection on the river shores and the land behind the dykes. To preserve the *bäuerliche Kulturlandschaft* agriculture is essential. Farmers have an important role in nature and landscape management in these areas.

Nature development is defined as measures aimed at the restoration of natural conditions, and the preservation and extension of remaining natural areas. Linking scattered nature areas is regarded as essential for the strengthening of these areas and the realisation of an ecological infrastructure (*Biotopverbund*). The (re)development of nature can contribute to this. Drastic nature development measures, by which a whole area is transformed into an 'new' kind of nature, are considered to result in secondhand nature (*Natur aus zweiter Hand*).

On the basis of a feasibility study (*Machbarkeitsstudium*) carried out in 1994, the Germans decided to continue with their existing policies (*Ist-Zustand*) and not to develop special measures for the *Gelderse Poort* area. Obviously the farmers in the area were relieved, because their role in nature management was acknowledged as important, if not as essential.

In the German discourse, nature in the *Gelderse Poort* is protected effectively through its designation as a Ramsar area and as a *Naturschutzgebiet* (nature conservation area) of about 5,000 hectares (of the altogether 11,000 hectares on the German side of the border) under federal and state legislation. Farming in *Naturschutzgebiete* is regulated in accordance with nature conservation goals. Farmers receive financial compensation for the adaptation of their management, settled in voluntary management agreements (*Vertragsnaturschutz*).

The countryside's own identity and potential is also emphasised with regard to tourism and outdoor recreation. Because of their impact on nature, proposals for tourism and leisure facilities must be co-ordinated and conform with policies for nature conservation. Nature protection areas can be important national or regional tourist attractions, which leads to a conflict between leisure and nature conservation. Indeed areas considered to be most ecologically sensitive are often the most intensively used tourism and recreation areas. The priorities are to provide recreation facilities for the local and regional population, while at the same time protecting nature and agriculture from potential damage by recreational use.

General policy arrangements were regarded as sufficient and adequate for the *Gelderse Poort* area. As a consequence of this perception and the assessment of current nature values, German policy on recreation differs from Dutch policy. In Germany the implementation of nature policy does not imply the withdrawal of land from farmers in order to 'give it back to nature'. Conservation goals can be reached through a combination of land use and function and in co-operation with present land use. The Germans pursue the integration of nature, the cultural landscape and agriculture. The reservation of specific areas for nature does not have to be legitimised to a broader public and therefore nature areas often are not open to the public or for outdoor recreation. Policy makers do not need the voluntary co-operation of private actors to implement their policy.

Dutch 'comprehensive nature development' coalition

In the Netherlands the majority of the actors involved in the *Gelderse Poort* network want to develop nature. Their definition of nature is associated with more or less ambitious and drastic developments and measures. The central story line in this discourse is the wish to expand the natural milieu along the river and to transform the agricultural land at the river foreland areas. This vision is based on the assumption that these areas were in a natural state many centuries ago. Wetland forests would have stood on currently treeless lands between the main dike and the main stream of the river (Mayr et al., 1996). Such a situation can even be duplicated under modern conditions, with a man-made regimen for the riverbed and modern agriculture behind the dykes. Separation of agriculture and nature is considered to be the right strategy to reach nature conservation goals. Examples of the spontaneous development of nature areas in neglected places, such as the *Oostvaardersplassen* (*Zuid-Flevoland*), provided a model for other regions. The conditions for nature development in the *Gelderse Poort* area are considered to be excellent. A complete river system, with elements of the river itself (open waters, muddy flatlands, river dunes, etc.) and the contact zones between the flood plain and the moraine hills, is supposed to be regenerated. This discourse is considered to be the most consistent implementation of the ideas in the NURG (Further Development Plan for the River Area).

Central actors in the coalition shaping these ideas are the Ministry of Housing, Spatial Planning and the Environment (*Ministerie van Volkshuisvesting, Ruimtelijke Ordening en Milieu*, VROM), the Ministry of Transport, Public Works and Water Management *(Ministerie van Verkeer en Waterstaat*, V&W) and the province of Gelderland. Representatives of the leisure sector, of clay extractors and of brick producers supported this vision at the start of the policy process in 1992.

The planned location and size of new nature areas are fundamental different with the other Dutch coalition (river nature, see below). The 'comprehensive nature development' coalition wanted to establish 3,000 hectares of new nature (mainly on former agricultural land) both inside and outside the dykes within fifteen years. The final goal is to develop large-scale interconnected pristine new nature areas.

The actors in this coalition support a relatively flexible regime with regard to tourism and outdoor recreation. The main argument is that, by giving people the possibility of experiencing the new nature, they will have more understanding of nature policy. In the Netherlands public support is important, because the feasibility of the policy hinges on the voluntary co-operation of many actors (farmers, landowners etc.). Farmers have to sell their land to the project organisation of the *Gelderse Poort* in order to let nature take over from agriculture.

Dutch 'river nature' coalition

The Dutch Ministry of Agriculture, Nature Management and Fisheries (*Ministerie van Landbouw, Natuurbeheer en Visserij*) together with the farmers formed a second Dutch coalition in the *Gelderse Poort* network. Its discourse differs in some essential respects from that of the previous coalition.

The Ministry of Agriculture wanted to use the ecological potential inside the dykes for nature development, with minimal human interference to create the essential conditions. The basic philosophy is to give nature itself a free hand on the river foreland area. Mainstream and intensive farming would be moved out of the flood-plain channels of the Rhine and its branches, and adapted farming by management agreement promoted. On this point there was a substantial difference of opinion between this coalition and the German one on the one hand and the comprehensive nature development coalition on the other. The 'river nature' coalition only wanted to develop nature at the river foreland area, while the other two groupings wanted dyke-crossing nature development. The area between the dykes is of less agricultural value and the river nature coalitions wanted to use the land behind the dykes for agriculture and give cultivated parts outside the dykes back to nature. The German coalition wanted to maintain their agriculture both behind the dykes and at the river foreland area and, together with the 'comprehensive nature development' coalition, pleaded for connected nature areas.

In the course of 1994 the province of Gelderland and the Dutch Ministries of Environment and of Transport changed their ideas about the location and size of new nature areas; all members of the Dutch steering committee reached agreement and became part of the same coalition (river nature). The dominant discourse in the comprehensive nature development coalition proved to be impracticable, because of the objections from farmers (who owned the land needed). It was not likely that the land required could be purchased from farmers. There simply was not enough land on the market and farmers did not intend to sell their land voluntarily. As a result parts of the areas behind the dykes were allocated to farmers who had to move out of the flood plains where nature development was planned.

The separation of nature and agriculture was more rigorously applied. The steering committee agreed upon a total surface area of approximately 2,000 hectares of new nature, thereby reducing the decrease in agricultural land by almost 1,000 hectares. The Ministry of Agriculture had argued for this from the outset of the planning process.

Facilitating tourism and outdoor recreation remained important in this coalition. Nature development gives (urban) people more room to experience nature. With the expansion of the towns of *Nijmegen* and *Arnhem*, the presence of green areas in the surrounding area is becoming more and more important. The type of primal nature (*oernatuur*) foreseen can withstand relatively intensive forms of recreational use and is freely accessible to everybody.

5.5 CONCLUSIONS

Looking back on eight years of policy making we can conclude that the grouping of coalitions in the *Gelderse Poort* is relatively stable. At the start of the policy process there were three dominant coalitions. The German coalition wanted to preserve the *status quo*. The central concept of the Dutch 'comprehensive nature development' coalition was the development of large-scale, interconnected, dyke-crossing nature areas (both inside and behind the dykes). The third coalition, 'river nature', had a more practical and feasible definition of nature and only wanted to develop nature in the flood plains outside the dykes. For several reasons the ideas of this coalition turned out to be preferable and politically the most practicable.

One of the decisive reasons was that in this vision it is not necessary to apply the same policy to the whole *Gelderse Poort* area. The German part is not considered to be supplementary to the Dutch (an extension of the same concept), but complementary to it. For the Dutch actors this was a great advantage, because they could develop their own national policy independently. Because of the strong resistance from farmers and the Ministry of Agriculture, Nature Management and Fisheries to the proposals to transform agricultural land behind the dykes into nature, the grouping in the coalitions changed around the end of 1994. The dominant discourse then was feasible for financial and political reasons. The ideas of this new coalition made it possible to obtain EU subsidies. The German-Dutch agreement provides a general framework in which both countries can work out their own ideas within a comprehensive, but general, transboundary policy arrangement. Additionally, developments in other policy domains, such as agriculture and the environment, changed the circumstances in rural areas and stimulated changes in nature conservation policy for the *Gelderse Poort*. The original ambitions of the comprehensive nature development coalition were modified in interaction with actors in the other two coalitions. In this sense a second important premise of the dominant discourse becomes clear.

Definitions of nature conservation differed at the beginning of the policy process and will continue to differ in the future. In the Netherlands nature development is associated with more or less ambitious and drastic developments. In Germany it signifies small changes in landscape and nature, whereby an existing nature area can be improved. Interactions in the context of the *Gelderse Poort* did not lead to a common discourse on nature conservation. The explanation for this is twofold. Firstly national policy arrangements are too different in Germany and the Netherlands. The institutional roots leave too little room for (lower-level) governments in regional projects to meet each other half way and to adapt their definition of reality to a shared policy. Secondly, in the case of the *Gelderse Poort*, transboundary interactions were insufficient, both with regard to the quality and to the frequency. The steering committee met two or three times a year in a quite formal setting. In these meetings officials and politicians had to make decisions on the basis of proposals from the project committees. Because of the few meetings and the formal setting actors did not really get to know each other. There simply was not enough time to give information on and learn about the points of departure in the two countries. Transboundary co-operation takes time and that was something the Dutch actors (notably the province) did not have. National policy arrangements dictated a more or less fixed time schedule, with insufficient time for transboundary policy formulation.

REFERENCES

Baerselman, F. and F. Vera (1989), *Nature Development, an exploratory study for the construction of ecological netwerks*, The Hague: Ministry of Agriculture and Fisheries.

Baldock, D., J. Jolzner and G. Bennet (1987), *The organisation of nature conservation in selected EC countries, a report on the nature conservancy council*, London.

Bennett, G. (1991), *Towards a European Environmental Policy*, Arnhem: Institute for European Environmental Policy.

Bennett, G. and R. Wolters (1996), A European ecological network, P. Nowicki, G. Bennett, D. Middleton, S. Rientjes and R. Wolters, eds., *Perspectives on ecological networks*: European Centre for Nature Conservation, pp.11-17.

Bolk, H. (1989), *Organizing, changing, simulating, an organizing perspective for management, consulting and science policy*, Delft: Eburon.

Cörvers, R. and W. Slot (1995), Forecasting land-use disputes; network analysis of a nature development project, P. Glasbergen, eds., *Managing environmental disputes: network management as an alternative*, Dordrecht: Kluwer Academic Publishers, pp. 137-153.

De Jong, D. (1999), *Tussen natuurontwikkeling en Landschaftschutz. Sociaal-cognitieve configuraties in het grensoverschrijdende natuurbeleid*, Delft: Eburon.

De Jong, D. and P. Leroy (1996), *Conflicting constructions of nature in transboundary policy on nature conservation and nature development*, the International Conference 'The Environment in the 21st Century: Environment, Long-Term Governability and Democracy, Abbaye de Fontevraud, France.

De Laat, W.A.M. (1983), *Vragen naar de onbekende weg: een sociaal-wetenschappelijk onderzoek naar deskundigheid in het kader van een theorie over organiseren*, Delft: Eburon.

De Swaan, A., eds. (1994), *A Social policy beyond borders, the social question in transnational perspective*, Amsterdam: Amsterdam University Press.

Felton, M. (1996), Natura 2000 the ecological network of the European Union: using buffer areas and corridors to reinforce areas designated by member states, P. Nowicki, G. Bennett, D. Middleton, S. Rientjes and R. Wolters, eds., *Perspectives on ecological networks*, Tilburg: European Centre for Nature Conservation, pp. 133-141.

Heidenreich, K. (1993), Grundsätze des Naturschutzes in Deutschland, Vorstellungen der Länderarbeitsgemeindschaft für Naturschutz, Landschaftspflege und Erholung (LANA), *Natur und Landschaft, Zeitschrift für Naturschutz, Landschaftspflege und Umweltschutz*, (heft 3), pp. 99-101.

In 't Veld, R.J., L. Schaap, C.J.A.M. Termeer and M.J.W. van Twist eds. (1991), *Autopoiesis and configuration theory: new approaches to societal steering*, Dordrecht: Kluwer Academic Publishers.

Jongman, R. (1995), Ecological networks in Europe, *Landschap*, (3), pp. 123-130.

Keulartz, J. (1997), De primitieve verbeelding aan de macht, Keulartz, J. en M. Korthals eds., *Museum aarde, natuur criterium of constructie?*, Amsterdam, Meppel: Boom, pp. 88-106

Klijn, E.H. and G.R. Teisman (1991), Effective policy making in a multi-actor setting: networks and steering, R.J. In 't Veld, L. Schaap, C.J.A.M. Termeer and M.J.W. Van Twist, eds., *Autopoiesis and configuration theory: new approaches to societal steering*, Dordrecht: Kluwer Academic Publishers, pp. 99-111.

Maas, A.J.J.A. (1988), *Ongedefinieerde ruimten: sociaal-symbolische configuraties*, Delft: Eburon.

Mayr, A. and J. Miggelbrink (1996), Cross-border problems of physical planning and ecological policy in the Dutch-German border region of Arnhem, Nijmegen, Kleve and Emmerich, H. Heineberg, N. De Lange and A.A. Mayr, eds., *The Rhine Valley,*

urban, harbour and industrial development and environmental problems, Leipzig, pp. 41-57.
Paterson, W.E. and D. Southern (1991), *Governing Germany*, Oxford, Cambridge: Blackwell.
Rokkan, S. and D.W. Urwin (1983), *Economy, territory, identity, politics of West-European peripheries*, London: Sage.
Roth, D. (1994), Zum Konflikt zwischen Landwirtschaft und Naturschutz sowie Lösungen für seine Überwindung, *Natur und Landschaft, Zeitschrift für Naturschutz, Landschaftsplfege und Umweltschutz*, pp. 407-411.
Schmidt-Eichstaedt, G. (1996), *The EU compendium of spatial planning systems and policies, Germany*, European Commission.
Stein, A.A. (1990), *Why nations cooperate, circumstances and choice in international relations*, Ithaca/London: Cornell University Press.
Stroming b.v. and Hydrologisch Adviesbureau Klink b.v. (1993), *Levende Rivieren, studie uitgevoerd in opdracht van het Wereld Natuur Fonds*, Zeist.
Termeer, C.J.A.M. (1993), *Dynamiek en inertie rondom mestbeleid: een studie naar veranderingsprocessen in het varkenshouderijnetwerk*, Den Haag: VUGA.
Van Houtum, H. (1998), *The development of cross-border economic relations*, Tilburg: Tilburg University Press.
Van Twist, M.J.W. and C.J.A.M. Termeer (1991), Introduction to configuration approach: a process theory for societal steering, R.J. In 't Veld, L. Schaap, C.J.A.M. Termeer and M.J.W. Van Twist, eds., *Autopoiesis and configuration theory: new approaches to societal steering*, Dordrecht: Kluwer Academic Publishers.
Van Vught, F.A. (1986), Op zoek naar een interactionistische sturingsconceptie: een poging tot theorievorming, M.A.P. Bovens, W. Derksen en W.J. Witteveen eds., *Sturing van de samenleving, conferentie over 'Het schip van staat'*, Zwolle, pp. 159-175
Voogt, A.A. (1990), *Managen in een meervoudige context, naar een methode voor het ontwikkelen en veranderen van sociaal cognitieve configuraties*, Delft: Eburon.

6. Global Environmental Policies: Between 'Interstatist' and 'Transnational' Arrangements

*BAS ARTS**

6.1 INTRODUCTION

The political modernisation theory as outlined in chapter 1 will be applied to *global* environmental policies in this chapter. Therefore it is rather different in nature to the other case studies (chapters 4, 5, 7 and 8). As a global focus will replace the national one, some introductionary comments on policies in the global system will be made in this introduction. These will be followed by a short history of global environmental policies (section 2). Such historical insights are needed in order to be able to diagnose policy change in the last three decades. But to do so, knowledge of current developments in global environmental policies is inevitably needed as well. Therefore sections 3 and 4, next, discuss two issues which arose in the 1980s and for which policies have been designed and implemented in the 1990s: biodiversity and climate change. By dealing with these empirical domains, data and insights will be gathered on the basis of which the political modernisation theory can be applied and, hopefully, enriched. Section 5 deals with the four dimensions of a policy arrangement - policy coalitions, power, rules, and discourses - and 'inserts' the empirical data of the earlier sections into this theoretical scheme. Subsequently, the political modernisation thesis - the stability and change of policy arrangements in time - is considered in section 6. In section 7, finally, some conclusions are drawn.

Before investigating the substance of global environmental policies, one should be aware of the fact that these are of a different nature to national policies. First of all, the international community lacks a central authority which has top-down legislative and authoritative powers (Lieshout, 1993; Waltz, 1979).

* The author wishes to thank Dr. Bob Reinalda, associate professor in international relations at the University of Nijmegen, for his comments on an outline version of this chapter.

Therefore global policies are mainly based on co-operative arrangements - agreements, declarations, treaties - between sovereign states (Hocking and Smith, 1990; Schermers, 1985a, 1985b). Secondly, environmental issues are less salient in the global arena than in for example Western Europe. International politics generally focuses mainly on security issues (so-called 'high politics') and less on environmental and other ones ('low politics') (Koch et al., 1994; Morgenthau, 1993). Consequently, agenda setting on environmental issues is less obvious at global than at national level. Thirdly, global policies are generally based on negotiations aiming for unanimity, as well as on consensus decision making. Consequently, common initiatives generally consist of broad and weak compromises. Fourthly, the national implementation of global policies is difficult to enforce, given the sovereignty of states (French, 1994; Liefferink et al., 1993). Moreover, there are hardly any sanctions with which to punish deviant nations, although states cannot simply ignore international commitments they agreed upon earlier, due to pressure from other states, Intergovernmental Organisations (IGOs) and Non-Governmental Organisations (NGOs) which are also involved in the agreement. However, compared to national ones, global policies are more difficult to design, implement and enforce.

6.2 GLOBAL ENVIRONMENTAL POLICIES: A SHORT HISTORY

It is generally assumed that the development of global environmental policies was initiated in 1972 when the United Nations Conference on the Human Environment (UNCHE) convened in Stockholm. However, UN organisations such as UNESCO, FAO and WHO had already been involved in the conservation of nature and of natural resources and the management of pollution before the 1970s (Caldwell, 1990; Caroll, 1988; Hurrell and Kingsbury, 1992; McCormick, 1989). Their initiatives were, however, sectoral in nature and it was not until the end of the 1960s that issues such as health, natural resources, energy and others were fused into one concept: 'the environment' (Jacobs, 1997). The UNCHE of 1972, which was the first major conference on this topic at global level, produced a declaration, an action plan, and a new, though small, UN body: the United Nations Environment Programme (UNEP). More significantly however, the UNCHE became the starting point for a large number of environmental initiatives at global level. Today, there are some 150 multilateral legal instruments on the environment, two-thirds of which were established after 1972; UNEP was involved in nearly half of these latter cases (Rowlands and Greene, 1992). Examples of such legal instruments are the Convention on International

Trade in Endangered Species of Wild Flora and Fauna (CITES), the Ramsar Convention on Wetlands of International Importance, the Basel Convention on the Control of Transboundary Movements of Hazardous Wastes, the Vienna Convention and Montreal Protocol for the Protection of the Ozone Layer, and the United Nations Framework Convention and Kyoto Protocol on Climate Change (Bergesen and Parmann, 1993; Caldwell, 1990)

Although the Stockholm conference marked the beginning of an era, public interest in environmental issues has been subject to fluctuations since then (Arts and Mulleneers, 1995; Weiss et al., 1994). The 1970s witnessed a decrease in public interest due to the oil crises, economic stagnation and high unemployment rates. Moreover, the doom scenarios of many environmentalists, which had been so popular at the end of the 1960s, were being increasingly criticised (Nelissen et al., 1997). As a result, the environment became a less prominent issue on the international agenda. Yet environmental politics continued to develop in the 1970s. Firstly, debates on environment and development - which were initiated by the developing countries - led to the adoption and elaboration of the concept of *eco-development* by UN bodies, principally UNEP. This notion refers to a development style which takes the fulfilment of basic needs by local communities as well as the maintenance of ecological and cultural diversity at the regional level as its starting-points (Sachs, 1980). Although it was an innovative development theory, it was more or less ignored by mainstream economics and politics, as it was not really incompatible with the dominant, large-scale, growth-oriented development styles of the capitalist and socialist economies in North and South at the time. Secondly, the 1970s witnessed the formulation and adoption of a great number of international environmental treaties. Hence, an international 'web' of environmental norms and regulations - or *regimes* - emerged in the 1970s (Young and Osherenko, 1993)

The environment reappeared unexpectedly high on the international agenda in the mid 1980s, for several reasons. Firstly, the world economy was doing better in this period than in the 1970s. Secondly, new environmental issues had arisen, such as global warming. Moreover, so-called shock events - the discovery of the hole in the ozone layer above Antarctica and the nuclear and industrial disasters in Chernobyl and Bhopal respectively - caused widespread concern. Thirdly, the easing of East-West tensions and the subsequent end to the Cold War created new opportunities in the environmental field (Opschoor, 1989; Käkönen, 1994). Finally, UN bodies undertook new initiatives.

In 1982 UNEP reviewed the Stockholm Conference of 1972. This review was not encouraging, given the low attention paid to environmental issues, so new initiatives were launched. The World Commission on Environment and

Development (WCED), headed by Gro Harlem Brundtland, was subsequently established and produced its famous report *Our Common Future* in 1987. This report urges governments to pursue *sustainable development*, that is, a development style which covers economic, social, ecological and long-term considerations on a global scale (WCED, 1987). Since its publication, sustainable development has become one of the buzzwords in the international community, especially as it is - in contrast to the eco-development approach of the 1970s - more compatible with mainstream economics and politics (Adams, 1991; Arts, 1994a; Lélé, 1991; Redclift, 1987). Furthermore, the Brundtland report foreshadowed the United Nations Conference on Environment and Development.

This UNCED (also called the Earth Summit) - held in Rio de Janeiro in June 1992 - produced the Rio Declaration on Environment and Development, the Authoritative Statement on Forest Principles, Agenda 21, and a new, though again small, UN body: the Commission on Sustainable Development (CSD) (Johnson, 1993). In addition, more than 150 countries signed the UN Framework Convention on Climate Change (FCCC) and the UNEP Convention on Biological Diversity (CBD), although these two instruments were designed in separate but parallel UN forums (sections 3 and 4). The Rio Declaration defines a number of legally non-binding principles, agreed upon by all nations (e.g. national sovereignty, global responsibility, precautionary approach and prevention). The Authoritative Statement on Forest Principles does the same with regard to the management of forests; and Agenda 21 - a non-binding action plan on sustainable development for the 21st century - deals with topics such as poverty, consumption patterns, natural resources, forests, waste, health, oceans and trade.

Recently, the *United Nations General Assembly's Special Session on UNCED* (UNGASS) was held in New York, in June 1997. This UN meeting assessed the implementation of Agenda 21 thus far, five years after the Earth Summit, but the conclusions were not encouraging. Journalists proclaimed this meeting a failure (ENB, 1997a). Alongside UNGASS, two additional aspects are important in global environmental policies in the late 1990s. Firstly, more emphasis is put upon policy implementation nowadays. It appeared that environmental regulations are more easily agreed upon than put into operation, something which is also referred to as the *implementation deficit* (Collins and Earnshaw, 1992; French, 1994). And secondly, public attention to environmental issues has decreased again, although not as dramatically as in the late 1970s and early 1980s. But the *Rio Hype* is definitely over.

Apart from states and international organisations, non-governmental organisations (NGOs) - in particular environmental pressure groups such as Greenpeace, Friends of the Earth and WWF - have played crucial roles in global environmental policies right from the start (Caldwell, 1990; Princen and Finger, 1994; Willets, 1982). They already pressurised the public and the politicians to address environmental matters in the early 1970s and thus contributed to agenda setting, also in the global arena. As UN rules enable them to observe and participate in UN meetings and conferences, NGOs were prominently present at the UNCHE and UNCED and were able to impact policy making through lobbying, advocacy and protest (Arts, 1998; Johnson, 1993; Princen and Finger, 1994; Thompson-Feraru, 1974). Other non-state actors have played roles also, in particular interest groups of transnational corporations (TNCs). The latter try to impact policies mainly through conference lobbies (Chatterjee and Finger, 1994; Mitchell, 1998).

Table 6.1 below summarises this section on the history of global environmental policies. It identifies five periods, each of which is characterised by specific aspects.

Period	Characteristics
- 1968	- sectoral initiatives by UN bodies (UNESCO, WHO)
1968-1972	- integration of issues through the notion of 'environment' - preparation for UNCHE
1972-1982	- expanding role of UNEP - formulation of environmental law - debate on 'eco-development'
1982-1992	- Brundtland commission and report (*Our Common Future*) - debate on 'sustainable development' - preparation for UNCED - expanding role of non-state actors (e.g. NGOs)
1992 -	- CSD and UNGASS - focus on implementation

Table 6.1: The development of global environmental policies

6.3 BIODIVERSITY: NEW CONCEPTS, NEW COALITIONS

It is generally acknowledged that the diversity of species on Earth is decreasing. For example, 20 per cent of all reptiles, 12 per cent of all flora, and 10 per cent of all mammals are endangered today (Glowka et al., 1994; Tolba and El-Kholy, 1992). Traditionally, this subject has been addressed at global level by sectoral conservation agreements. Examples are the Ramsar Convention on Wetlands of International Importance (1971) and the Bonn Convention on the Conservation of Migratory Species of Wild Animals (1979) (Bergesen and Parmann, 1993). These agreements were (1) mainly based on biologists' and conservation organisations' arguments (e.g. Huxley, Leaky, Myers, IUCN, WWF); (2) developed within the context of international organisations (particularly the UN); and (3) endorsed by the wealthiest nations (e.g. USA, UK, France) (Caldwell, 1990; McCormick, 1989).

One of the main objectives of these agreements was to protect nature from adverse human intervention and a key policy instrument used was the establishment of protected areas - or 'nature parks' - around the globe. Human activities were banned from such valuable sites, except for tourism or some limited exploitation (small-scale logging, hunting, fishing, farming, etc.). However, the current number of protected areas around the world is, according to biologists, too small to safeguard the diversity of life on earth (Glowka et al., 1994; WCED, 1987). Also, current agreements are, according to lawyers, insufficient to adequately conserve nature (Sanchez and Juma, 1994). Besides, the concept of protected areas puts natural environments outside parks beyond the law. It has also provoked many conflicts between conservation groups on the one hand and so-called 'user groups', such as local farmers, logging companies and even governments of developing countries (which wish to *develop* rural areas instead of *conserving* them, on the other). These disadvantages, particularly acknowledged in the early 1980s, impelled conservation groups and governments to look for new policies to enhance the protection of nature.

In 1984 the World Conservation Union (IUCN) - an NGO - proposed the developing of one global instrument to conserve the diversity of life on earth, including genetic diversity, to overcome the shortcomings of nature conservation policies at that time (Burhenne-Guilmin and Casey-Lefkowitz, 1992). IUCN commissions prepared drafts for a convention and, for the first time, the notion of *biological diversity* was introduced. This notion means (UNEP, 1994: 4): 'The variability among living organisms from all sources, including, *inter alia*, terrestrial, marine and other aquatic ecosystems and the ecological complexes of which they are part; this includes diversity within species, between species and of ecosystems'. According to this definition, biological diversity brings together different areas of concern. Firstly, the problems relating to genetic erosion and genetic engineering, secondly the endangering and extinction of species and thirdly the destruction and loss of ecosystems. Hence, the concept of biological diversity is much broader than the traditional concept of nature conservation. Its introduction is therefore the first discursive innovation in this field in the 1980s.

In 1987, the United Nations Environment Programme (UNEP) - which is, in contrast to the non-governmental IUCN, an *intergovernmental* body - took up the matter of a global convention on biological diversity. Experts' Working Groups were established, which advised the UNEP to start formal negotiations on a Convention on Biological Diversity (CBD). The Intergovernmental Negotiation Committee (INC) began its work in 1990 and in 1992 the text of the treaty was adopted and signed by more than 140 nations. Given its objective in Article 1, the CBD addresses many aspects of biodiversity. 'The objectives of this Convention (...) are the conservation of biological diversity, the sustainable use of its components and the fair and equitable sharing of the benefits arising out of the utilization of genetic resources, including by appropriate access to genetic resources and by appropriate transfer of relevant technologies (...) and by appropriate funding (UNEP, 1994: 4). Following the introduction of the concept of biological diversity, the emphasis on *sustainable use* of biodiversity and the *sharing of its benefits* is a second innovation in international nature conservation policies (Sanchez and Juma, 1994). Here the concerns and interests of both conservation and user groups have been integrated. This innovation was due to effective lobbying by the group of developing countries (G77), in this case supported by the Nordics - Sweden, Denmark, Norway, Finland - and some NGOs (Arts, 1994b; Miller, 1995). The G77 was not willing to support a biodiversity treaty which would not address their development needs. It also wanted to stop the free extraction and patenting of natural resources and genes within their territories by Western multinationals and research institutes (Juma, 1989).

A third innovation in international conservation policies concerns the involvement of NGOs and women's groups as well as indigenous and local communities in the CBD. This is, incidentally, not surprising, since some of these global, national and local groups - and ever more of them since 1991 - have been part of the drafting and implementation process of the convention. To some extent, these groups were able to influence the text and the initial implementation of the CBD (Arts, 1998). A final renewal in policy making is the abolition by the CBD of the protected area approach. After all, the conservation and sustainable use of biological diversity should, according to this treaty, happen *everywhere*, not only in special areas.

To conclude, the shift from nature conservation to biodiversity policies, which took place in the 1980s and 1990s, is characterised by three innovations: firstly, the notion of 'biodiversity' itself; secondly the integration of conservation and sustainable use; and thirdly the involvement of new and more private (and even local) groups. However, it remains to be seen whether the CBD will become an effective instrument in renewing global conservation policies in the near future.

6.4 CLIMATE CHANGE: NEW POLICY INSTRUMENTS

The so-called greenhouse effect was already understood more than a century ago (Mintzer and Leonard, 1994; Schneider, 1989). Some gases in the higher atmosphere - including water vapour (H_2O), carbon dioxide (CO_2), nitrous oxide (N_2O), methane (CH_4), ozone (O_3) and man-made chlorofluorocarbons (CFCs) - are transparent to the incoming short-wave solar radiation, but absorb infrared radiation from the planet's surface. As a consequence, these gases warm up the atmosphere, which is - by the way - a natural *and* a necessary process to sustain the Earth's climate. However, due to additional greenhouse gases produced by humans - particularly CO_2 emitted by industry and transport - the greenhouse effect is likely to be strengthened. This phenomenon is referred to as *global warming*. The Intergovernmental Panel on Climate Change predicts an increase in global mean temperature in the range of 1.5 to 4.5° C during the next century (IPCC, 1990a). The possible impacts could be disastrous for many regions of the world: more droughts in tropical regions and more rain in moderate ones, less output in agriculture and forestry, the disappearance of several species and habitats, the spread of malaria, and the melting of snow, ice and permafrost which will cause a rise in sea level (IPCC, 1990b). Policy responses are therefore needed.

On 21 December 1990, the UN General Assembly established the Intergovernmental Negotiation Committee for a Climate Convention. This INC held six sessions in the period 1991-92 and adopted the Framework Convention on Climate Change (FCCC) in May 1992 (Mintzer and Leonard, 1994). At the UNCED, it was signed by more than 150 countries. The most controversial issue of the treaty and its implementation is *targets and timetables*. This phrase refers to the dominant policy instrument of atmospheric politics since the 1970s (Bodansky, 1993). The instrument aims to bind individual states to general and common emission reduction targets for 'pollutants' (such as SO_2 in the case of acid rain, CFCs in the case of the protection of the ozone layer and CO_2 in the case of global warming). However, views on the severity and scope of a specific target and timetable for greenhouse gases (GHGs) differed widely. Whereas the IPCC, European countries and 'victim states' - e.g. small island states in the Caribbean and Pacific - proposed relatively strong targets for GHGs, the 'great emitters' - such as the USA and Japan - as well as the oil-exporting countries (OPEC) opposed these. This led to months of complex crisis diplomacy within and outside the INC. Consequently, the FCCC contains only a weak and vaguely formulated commitment from the industrialised world to stabilise its GHG emissions by the year 2000 at 1990 levels (so-called *stabilisation target*). Additionally, the parties to the FCCC agreed at the Kyoto Conference in Japan, in December 1997, to adopt a protocol to the FCCC to cover the period after 2000. This says that industrialised countries should reduce their emissions by 6 to 8 per cent compared to 1990 levels between the years 2008 and 2012 (ENB, 1997a).

One of the main challenges for the FCCC and the Kyoto Protocol is the implementation of the agreed targets and timetables, especially since the emissions of GHGs of most industrialised countries are still *increasing* (INC, 1994). Hence, the stabilisation target and Kyoto targets will probably not be attained. Nonetheless, many developing states and environmental organisations maintain their claim that industrialised countries simply *have* to comply with the targets and timetables they have accepted and, therefore, that they need to introduce tough policy measures. The industrialised countries and economic interest groups, though, tend to choose another option: the introduction of so-called *flexible instruments* (ENB, 1997b).

The phrase 'flexible instruments' refers to a set of innovative policy mechanisms to reduce GHG emissions other than through a general and common target and timetable which is imposed on all countries. Examples are Joint Implementation, Clean Development Mechanism, Tradeable Emission Permits and Benchmarking (Duijse et al., 1998; ENB, 1997b; Woerdman, 1997). The instruments *Joint Implementation* and *Clean Development Mechanism* are based on the premise that - given the fact that climate change is a global problem - it does not matter much *where* on earth GHG emissions are reduced. Hence, one should do it where it is cheapest. The more the marginal cost of new investments in cleaner technologies increases, the more efficient and environmentally sound an economy is. One should focus on investment in Eastern Europe and the Third World to reduce GHGs. Joint Implementation (the instrument for Eastern Europe) and Clean Development Mechanism (the instrument for the Third World) therefore imply that the industrialised countries implement the provisions of the FCCC and Kyoto Protocol jointly with the less efficient former communist states and with developing countries. In fact, the 'rich' provide money and technology, which is invested in the economies of the 'poor' in order to decrease emissions. As a consequence, industrialised countries may achieve their targets abroad and do not need to restructure their domestic economies.

A system of *Tradeable Emission Permits* allows countries to buy and sell permits to emit GHGs. A central authority releases these permits and their number is related to a certain ceiling on total emissions. Subsequently, each country receives a number of permits. The most energy-efficient ones probably do not need all of them, as they emit relatively fewer GHGs, and so they may sell some of their permits on the emission trading market. The opposite applies to energy-intensive countries, which probably need more permits. Either they can buy these in the market - which costs a great deal of money - or they can decide to improve their performance on GHG emissions. There is therefore an inherent incentive in this system to improve environmental performance and, consequently, the total quantity of GHG emissions will be reduced in time.

Benchmarking, finally, is a mechanism to improve energy efficiency within industrial sectors and, therefore, to reduce GHG emissions. In short, companies in one sector - under supervision of a government or an international organisation - commit themselves to (incremental) targets on efficiency improvement (self-governance). Although these targets are not legally binding, proponents believe that they provide incentives for corporations to improve their performance and, hence, emit fewer GHGs.

It should be realised, however, that these flexible instruments have barely been implemented yet and, moreover, are still being criticised in the climate arena. Critics believe that these instruments were only invented to undermine the FCCC and help the great emitters of GHGs to escape from their obligations. After all, flexibility may enable industrial countries and companies to transfer their obligations to others or to the future. Given this criticism, the new instruments are now under discussion. This discussion, nonetheless, marks the beginning of greater influence and input from the private sector in the climate arena. State-oriented instruments, such as legally binding targets and timetables, seem to lose ground in favour of market-oriented instruments, such as joint implementation, emission trading and benchmarking.

6.5 POLICY ARRANGEMENTS AT GLOBAL LEVEL

In the above, developments and new trends in global environmental policy making and implementation have been dealt with, and the changes in actors, policy instruments and concepts in environmental politics highlighted. The question now is how these phenomena can be interpreted in terms of the political modernisation theory central to this book. How did in fact the organisation and content of global environmental policies changed in the last three decades? Can one speak of change in policy arrangements in the global arena? And might old and new arrangements co-exist today? These questions will be addressed below. This section presents the developments with regard to the four dimensions of a policy arrangement: policy coalitions, power, rules, and discourses. In section 6.6 these insights will be related to the political modernisation thesis.

6.5.1 The Broadening of Policy Coalitions

Although the study of international relations and in particular the mainstream of *(neo)realism* tends to focus on the nation state as 'the' relevant actor in international politics (Morgenthau, 1993; Waltz, 1979), recent literature distinguishes the following important players: states, intergovernmental organisations (IGOs), transnational corporations (TNCs), non-governmental organisations (NGOs) and epistemic communities, the latter being networks of experts on a specific policy issue (Haas, 1993; Hocking and Smith, 1990; Jacobson, 1984; Ray, 1987; Strange, 1988; Waters, 1995; Willets, 1982). In many cases of policy making and implementation, these players tend to operate in more or less stable policy coalitions.

The most obvious and classic example of coalition formation in international politics is that among states. States tend to operate in (regional) blocks, in international environmental politics as well. One may think for example of the G77 and China (group of developing countries), OPEC (Oil Producing and Exporting Countries), the AOSIS (Alliance of Small Island States), the OECD (the Western industrialised countries), the EU (Western Europe) and the CEE (Central and Eastern Europe). Given similarities in development level, regional backgrounds, culture and/or environmental conditions, these countries share specific interests, which they may jointly defend in the global arena. As the objectives of such policy coalitions can differ substantially, many confrontations on global environmental policies have occurred. A common divide is the one between North and South (or the OECD and the G77 respectively). Whereas the rich countries have favoured at least some regime formation on the environment since the 1970s, the poor ones were reluctant to do so, as they focused on economic development and feared that environmental protection would hamper that process (Miller, 1995; Redclift, 1987). But other confrontations have been common as well, for example between the EU and the USA. Generally, the EU is more progressive on environmental matters than the USA, as the climate issue has shown.

Although intergovernmental organisations (IGOs), the first group of so-called *non-state actors*, generally function like international *arenas* in which states negotiate and decide on policies, they may also operate as *players* themselves (Archer, 1983; Arts and Mulleneers, 1995). Examples are the FAO, UNEP, World Bank, UNESCO, WMO and WHO, representatives of which were involved in the formation of the Climate and Biodiversity Conventions. Additionally, these organisations may co-operate with other actors. For example UNEP works closely together with individual state representatives, experts and NGOs to strengthen international environmental regime formation.

The three other groups of non-state actors - TNCs, NGOs and epistemic communities - may form policy coalitions on their own (although their status as 'non-state' depends on how they relate to or are subject to governments; in particular experts are often bound to governments). TNCs themselves are generally absent in international political arenas, but their interests are furthered by sectional groups (e.g. the oil and coal lobby in the climate arena) and, of course, by 'anticipating' policy makers (see next subsection). Moreover, their direct input seems to have expanded recently, as the involvement of companies in the climate arena shows. Non-profit NGOs such as environmental pressure groups are generally better represented and more 'visible' at international negotiation sessions, at least since the late 1980s. They also try to co-ordinate their input through coalitions. But this goal is not always attained. Whereas the Climate Action Network (CAN) functions as a very successful coalition of NGOs in the climate arena, the biodiversity arena lacks such a broad NGO network. Epistemic communities, finally, are a rather distinct group, as they do not represent specific economic or environmental interests, as respectively TNCs and NGOs do, but are supposed to offer 'objective' expertise. However, experts are more often than not, directly or indirectly, associated with states, IGOs, NGOs and/or TNCs, so that they are not just providers of neutral knowledge, but involved in policy coalitions as well. Besides, the role of expertise has become crucial to international environmental policy making in the 1990s - given the complex nature of, for example, global warming - so that epistemic communities have become very important players (Boehmer-Christiansen, 1994).

Although all groups of state and non-state actors may form group-specific coalitions, cross-sectional ones have become common in international environmental politics as well. Traditionally, cross-sectional policy coalitions have been organised around economic interests - both 'rich' and 'poor' - as well as around environmental interests (Sabatier, 1987). Thus, three classical policy coalitions have been created in global environmental politics: (1) a *developmental* coalition of developing countries, IGOs (e.g. UNDP), and developmental NGOs and experts; (2) an *economy-oriented* coalition of 'conservative' industrialised countries, IGOs (e.g. OECD), the business lobby and economists; and (3) an *environmental* coalition of industrialised 'progressive' countries, IGOs such as the UNEP and environmental NGOs and experts (e.g. ecologists). However, this classification of policy coalitions is to some extent imprecise and dated. Imprecise, because coalitions may vary with each specific issue, and dated, because traditional borderlines between organised interests seem less relevant today. For example, Northern and Southern countries did recently co-operate on certain issues and, in that way, transcended the traditional North-South divide (e.g. EU and AOSIS in the climate arena), and even business groups and environmental NGOs, which seemed to represent opposite interests so far, have done so. A good example is that of Greenpeace and the insurance industry, which in the climate arena jointly favoured strong reduction targets on greenhouse gases to prevent global warming and, hence, hurricanes, floods, bad harvests etc.

Transcending traditional policy coalitions is one aspect, the input of ever more *national* and *local* players in international policy coalitions another. The case of the Biodiversity Convention is an illustration in point. Through international NGOs such as Greenpeace and the Third World Network, and later through their own international networks, local farmers and indigenous people were heard in the biodiversity arena at global level. They told the delegates about the effects of the loss and patenting of 'their' biological diversity. This seemed to have an astonishing impact - bringing local interests to global political decision making - and, consequently, this phenomenon has become rather more common in current international environmental politics (Princen and Finger, 1994).

To bring this subsection to a conclusion, since the 1980s four trends in policy coalition formation can be distinguished. Firstly, the involvement of ever more non-state actors (IGOs, NGOs, TNCs, epistemic communities), secondly the formation of coalitions which transcend traditional boundaries (e.g. North joins South; NGOs join TNCs), thirdly the strengthening of local-global links in policy coalitions (e.g. participation of local farmers and indigenous people) and fourthly the increase in flexibility in policy coalition formation (as a general effect of the former three trends). This may be summarised as the *broadening of policy coalitions*. However, this does not imply that the old, traditional and relatively stable policy coalitions of international environmental politics - the developmental, economy-oriented and environment ones - have become obsolete. On the contrary, traditional conflicts may still arise and put the old patterns of coalition formation into operation. This is also dependent on the issue concerned. An example is forests, an issue which so far has led to ongoing clashes between traditional interests. On other issues, though, (some of) the old contradictions could be overcome, so that new coalitions could emerge. Biodiversity is a good example in this regard.

6.5.2 The diffusion of power

International arenas in which environmental policies are designed and reviewed are also systems of power. To some extent, these resemble the general (uneven) power structure of the international system, in which - at least according to realists - the biggest nations determine outcomes. This claim is challenged by others (Guzzini, 1993). Neo-Marxists, for example, believe that today's TNCs are as powerful as states, while transnationalists believe that IGOs and NGOs may be powerful in specific cases of policy making as well (Keohane and Nye, 1989; Strange, 1988). These differences are also partly due to divergent views on the concept of power. Realists focus on the general and stable *power structure* of the state system as a whole, while others merely focus on specific policy *processes* in which actors other than states may be *influential* (Clegg, 1989; Guzzini, 1993). From the perspective of policy arrangements, both concepts may be relevant. The concept of power structure refers to a more or less stable division of allocative and authoritative resources in a policy arrangement, the concept of political influence to the effects of the players on specific policy outcomes within that arrangement. These two aspects are of course related, but there is no one-to-one relationship in the sense that those who possess most resources will always most affect policy outcomes (Keohane and Nye, 1989).

If we look at the recent history of the global system, the most obvious event is the fall of the communist bloc in the late 1980s (Koch et al., 1994). As a consequence, the system changed from a bipolar power structure into one dominated by the USA. This transition has affected all policy domains and arenas in international relations, including those on the environment. These events pose the question whether the USA may and can act as a 'hegemon' in the environmental field today. In other words, does the USA more or less on its own determine the outcomes of international environmental agreements? Some indeed believe so and refer, for example, to the strength of the Montreal Protocol and the weakness of the Climate Convention (Kaelberer, 1992). Both outcomes are, according to these authors, due to the USA's influence and in accordance with its interests. Others, though, believe that this analysis is biased (Paterson, 1996). They consider that the success of the Montreal Protocol was also due to UNEP and NASA as well as some NGOs and TNCs. And although the FCCC and Kyoto Protocol might indeed be called weak, the agreed obligations still exceed those initially wished for by the USA. What is more, the international community decided to accept a Biodiversity Convention in 1992, with which the USA could not live at all. It therefore decided not to ratify this convention (so far). So international agreements do not always match the USA's interests.

Alongside the USA, other regional powers affect international environmental politics. One might think here of the EU and Japan, but also of Brazil, China and India. Although many believe that the Northern countries dominate those of the South in international policy arenas, others have discovered that the G77 and China were very well able to fundamentally impact the content of international environmental policies (Arts, 1994b; Miller, 1995). The Biodiversity Convention is - again - a good example. As section 6.3 has shown, this treaty was originally designed to be a nature conservation treaty, but the developing countries were able to redefine its premises in accordance with developmental considerations. Although a power structure in terms of resources might be up against a player or a group of players - USA versus the rest, North versus South - the latter may still be able to affect policy outcomes. This is even true for smaller countries and non-state actors. Countries such as Nauru and the Bahamas are of course weak and tiny, but their possible fate of being one of the first victims of global warming has made them relatively influential actors in the climate arena.

The same applies to a number of NGOs (Arts, 1998). An assessment of their political influence revealed that they were able to affect several political decisions on the climate and biodiversity issues (e.g. on joint implementation). The impact of other non-state actors on the climate debate was also obvious. The epistemic community in the global warming field - of which the IPCC is the main exponent - has been tremendously influential in putting the issue on the political agenda and in pressing for strong regulations (Boehmer-Christiansen, 1994). Additionally, the business community has become a major player, for example through the *Global Climate Coalition*, in which many multinationals in the coal and oil industry participate (Mintzer and Leonard, 1994; Mitchell, 1998). It was found that the business lobby played a role in diluting the final outcomes at Rio (stabilisation target) and Kyoto (6-8 per cent reduction target). Besides, it should be remembered that business interests are generally more easily anticipated by political decision makers than environmental interests. Therefore companies may be influential even without being active lobbyists (Lukes, 1974).

The above analysis reveals that, although the current power structure of international politics suggests dominance by the USA, political outcomes in environmental arenas are determined by a wide range of players. In other words, political influence in international environmental arenas is (unevenly) *diffused* today. This observation raises the question whether one can also speak of a *process of diffusion*, which would imply the thesis that political influence is more dispersed today than it was ten or twenty years ago. It seems that this question may be positively answered for the following three reasons. Firstly, the number of players in the global environmental arena has definitely increased, as the previous subsection has shown. This phenomenon *in itself* has already induced a process of diffusion of political influence. Secondly, since the 1960s the state as a 'closed' power container and as the ultimate body of political authority has lost its exclusiveness (Beck, 1997; Giddens, 1984; Willets, 1982). This has led to the rise of other players with political legitimacy (United Nations, Greenpeace, Amnesty International, IPCC etc.). The division of authoritative resources has therefore changed in favour of non-state actors. Although there is no one-to-one relationship between resources and influence, this new power structure has definitely led to more opportunities for non-state actors to impact policy outcomes. Finally, the end of the Cold War has also opened up new windows of opportunity. The so-called 'high issue' of military security no longer hijacks political agendas and players. There seems to be more room for manoeuvre today, which has also enabled non-state actors to achieve more of their goals.

To conclude this subsection, there seems to have been a (relative) change in both power structure and the division of political influence in the global environmental arena since the 1970s. The political legitimacy of non-state actors has been strengthened, while their impact on policy outcomes has increased at the same time. This process can be described as the *diffusion of power* in environmental policy arrangements. However, one should put this process into perspective, as players such as the USA and EU remain dominant in terms of resources and political influence *vis à vis* others.

6.5.3 The 'pluralisation' of rules

International environmental policies at UN level are, of course, made in accordance with UN rules. These rules, of which those of the UN Charter are the most important, are based on the premises of common interests on the one hand (peace, security, development, international community) and the sovereign state on the other (the key to membership and decision-making power) (Kaufmann, 1980). At the same time, Article 71 of the Charter states that the Social and Economic Council (ECOSOC) may consult international NGOs - promotional, scientific and economic interests groups - which are concerned with UN matters, and give them accreditation to ECOSOC sessions as well as UN conferences on special issues (Feld and Jordan, 1983; Jacobson, 1984; Willets, 1981). Accredited NGOs have observer status at such meetings, may make oral statements, and can disseminate written position papers among governments. In addition, they may have access to working groups in which negotiations between countries take place, but the officials decide this on a case-by-case basis. Countries may also add NGO representatives to their delegations at international meetings, but this is dependent on the willingness of individual countries and is not yet common practice.

The foregoing makes clear that UN rules on policy making were to some extent pluralist in nature right from the start. However, for as long time the rules on access remained rather restrictive and, moreover, they varied on several occasions. To be able to participate, NGOs needed to have consultative status with the ECOSOC, which was not a status achievable by all, especially as an organisation had to match the UN criteria and had to have enough resources to participate in global meetings. Besides, many states have been reluctant to accept NGOs as observers. Examples of such states are a number of African and Asian countries, because they are concerned that Islamic fundamentalists - masquerading as NGOs - would interfere in UN matters or that human rights groups would be too critical of them at UN meetings. Nonetheless, the rules have been less tight in the last few years, in particular since the UNCED, which is considered a watershed in the participation of NGOs in UN matters (see section 6.2). Any NGO which wishes to participate and is considered a relevant contributor by UN officials can become accredited to any meeting today, irrespective of whether the organisation concerned has consultative status with the ECOSOC or whether the NGO is international, national or local in nature. Additionally, the criteria for achieving consultative status with the ECOSOC have been broadened recently (ECOSOC, 1995). In short, the rules of the game have changed in favour of the participation of non-state players in UN affairs, although they do not have any *formal* negotiating role or any *formal* decision-making power so far (Willets, 1996). Informally, though, NGOs *do* play a negotiating role and influence decision making. For example, NGO representatives headed intergovernmental working groups in the UNGASS, the special meeting of the General Assembly on the aftermath of the UNCED in June 1997 (ENB, 1997).

So far the (in)formal *arena* rules have been discussed. These co-determine the access and interaction of players in political arenas. Alongside these, however, more comprehensive and covert rules are in operation in any political system: the so-called *meta* rules. These co-determine how (global) politics is signified and legitimised in the first place (Giddens, 1984; Wendt, 1987, 1992). Traditionally, international politics has been signified and legitimised merely through nation states, political leaders, their armies and their diplomats (*interstatist* system). Politics and war were considered as two sides of the same coin and the nation state - with its monopoly on legitimate violence - was the ultimate locus of political power (Morgenthau, 1993). In the postwar, nuclear age, though, maintaining this image of international politics would have led to the nuclear winter, so a security regime was introduced based on the UN system (*intergovernmental* system) (Koch et al., 1994). A balance of power between the USA and USSR was sought not only in terms of weaponry, but also through international organisations, especially the UN and its Security Council. Now war was no longer an option in foreign politics, but common security was (although several regional conflicts were fought, for example in Vietnam and Afghanistan) (Ray, 1987). Also, topics other than peace and security appeared on the international agenda: economic development, social issues, human rights, the environment etc. (Hocking and Smith, 1990). These topics are not so much linked to the nation state as a political institution, as to (sectors of) human society. Consequently, stakeholders from society have begun to participate in global politics more regularly *themselves*. Therefore the nature of the system may be transformed from one based on the state as the sole player into one referring to a wide range of actors. This change may lead to a situation in which global politics will be 'automatically' signified and legitimised in terms of a 'plurality'. Then the involvement of non-state actors will no longer be questioned, but will be considered 'normal' (*transnational* system).

To conclude this subsection, the rules of the game in global politics - both arena and meta rules - seem to have been (slightly) changed in favour of non-state actors. This may be labelled the *'pluralisation' of rules*. This does not imply, however, that the nation state has lost its privileged position in global politics. On the contrary, it is still the ultimate locus of *formal* decision-making power, but today it has to share the global arena with ever more other players.

6.5.4 The integration of discourse

The ideas, concepts, and categorisations on the environment in the international arena have evolved in a specific way. A number of phases can be distinguished. Firstly, several issues - e.g. health, water quality, sewage treatment, industrial waste, energy - were brought under the one umbrella of 'the environment' in the late 1960s (Jacobs, 1997). This was due to the intervention of scientists, activists and politicians. At global level, the government of Sweden played an entrepreneurial in this respect (McCormick, 1989). It brought the acidification of Swedish lakes to the attention of the General Assembly of the United Nations in 1967 and put it in a broader perspective of ecological interconnectedness, as the sulphur dioxide causing the problem was emitted in England. Sweden therefore stressed the need to address environmental problems at UN level. Subsequently, the UN decided to organise a global conference on the environment (UNCHE, Stockholm, 1972). Secondly, under pressure from the developing countries, specialised UN organisations and NGOs, the issue of the environment was linked to the issue of development in the early 1970s (Redclift, 1987). These countries and organisations feared that the development process in the Southern nations would be hampered if the international community imposed high environmental standards upon them. The linkage of environmental norms to developmental needs would prevent this happening. One of the results of this linkage was the concept of *eco-development*, which came into being in the late 1970s, although it remained outside mainstream thinking (see section 6.2). Thirdly, the 'environment and development debate' was expanded in the concept of *sustainable development* during the late 1980s and early 1990s. This time mainstream economists and politicians were participants in the debate as well. Furthermore, political and social issues were more prominent than ever before.

If the above analysis is valid, then the evolution of the environmental debate at global level can be summarised as the *integration of discourse*. After all, ever more topics were brought together under one umbrella: firstly, different material problems under the notion of 'the environment'; secondly, environmental and developmental issues under the notion of 'eco-development'; and thirdly, ecological, economic and social aspects under the notion of 'sustainable development'. Furthermore, this last concept has enabled the integration of heterodox and mainstream thinking. Nowadays, neo-liberal economists and environmental activists sit at the same negotiating tables, something which was unthinkable in the 1970s and early 1980s (Mitchel, 1998). This is not to say that they all agree, as several interpretations of sustainability do exist (Arts, 1994a). But the concept of sustainable development as such is broadly accepted, which was not the case with the notion of 'eco-development' in the 1970s. Therefore it is possible to speak of an integrative, consensual discourse.

Not only has the environmental discourse itself undergone some change, the discourse on controlling environmental problems has as well. As the climate case has shown (section 6.4), market-oriented instruments are gaining potency *vis à vis* state-initiated instruments. No longer is it thought that command-and-control policies of state institutions and international organisations will suffice to reduce environmental problems; self-governance by stakeholders and measures which fit in with the logic of the market place are needed as well. This process may be labelled integration of (governance) discourse too, because *both* state-oriented and market-oriented policy instruments are considered relevant in addressing environmental issues today.

The question now is how these substantive developments in global environmental policies relate to the organisational and institutional changes described in the subsections 6.2 to 6.4. On the one hand, it is more than likely that the input from new actors in political arenas enriched the substance on which discussions were held (*organisation affecting substance*). For example, the growing influence of non-profit NGOs has probably strengthened the ecological dimension of sustainable development in international environmental policies. What is more, the notion of *biodiversity* would not have existed as a policy concept *without* such NGOs. At the same time the input from economic interest groups has strengthened the economic dimension of sustainability, as well as the legitimacy of market instruments in solving environmental problems. On the other hand, the acceptance of new ideas and concepts by governments has enabled new players to become involved (*substance affecting organisation*). For example, the growing popularity of 'sustainable development' has opened up new windows of opportunity for both promotional pressure groups and economic interest groups to participate in global arenas.

To conclude, discourses on global environmental problems and their control have changed substantially over the last three decades, a change which can be summarised as an integration of economic, political, ecological and social ideas, concepts and categorisations. Additionally, there is more consensus now than in the 1970s on how to define and manage problems. Again, one should not overestimate these trends, as contradictions and conflicts between the parties continue to exist. To some extent, these conflicts are hidden by those same broad, integrated, and consensual discourses (Adams, 1991).

6.6 POLITICAL MODERNISATION AT GLOBAL LEVEL

In the previous section, developments in global environmental policies have been summarised as follows: the broadening of coalitions; the diffusion of power; the pluralisation of rules; and the integration of discourse. Given these changes concerning the *dimensions* of policy arrangements, the arrangements *themselves* have probably changed as well. However, there is a problem of categorisation. The ideal-types that were distinguished in chapter 2 - (neo)statist, (neo)liberal and (neo)corporatist arrangements - are not applicable at global level, as these labels are deduced from politics *within* the context of nation states. An alternative categorisation, which takes the nature of global policies, the international relations literature, as well as the analysis in section 6.5 into account, seems more appropriate for use in this chapter. In table 6.2, three ideal-types of policy arrangements are distinguished: interstatist, intergovernmental and transnational arrangements (a typology derived from amongst others: Archer, 1983; Hocking and Smith, 1990; Hogenboom, 1998; Jacobson, 1984; Reinalda, 1997; Risse-

Kappen, 1995; Waltz, 1979). An *interstatist* arrangement is characterised by state dominance and, therefore, by disagreements between (regional blocks of) states. This is expressed through a dominance of regional policy coalitions, hierarchical power relations, 'traditional' rules of the game (national sovereignty!), and confrontational policy discourses. A *transnational* arrangement - the other extreme - is characterised by flexible coalitions of governmental and non-governmental stakeholders, global-local linkages, diffusion of power, rules based on the political legitimacy of stakeholders, and by integrative discourses. An *intergovernmental* arrangement, finally, is somewhere in between. States are still dominant, just as in the interstate arrangement, but IGOs play more important roles, as do NGOs, TNCs and epistemic communities. It should be noted, though, that these arrangements do not exist in reality, but are ideal-types. Therefore, concrete policy domains may exhibit more or less interstate, intergovernmental and/or transnational characteristics.

	Interstatist Arrangement	**Intergovernmental Arrangement**	**Transnational Arrangement**
Coalitions	Regional blocks of states (East versus West; North versus South)	Regional blocks of states; also coalitions of non-state actors	Flexible coalitions of stakeholders; local-global linkages
Power	Dominance of (big) states	Dominance of (big) states; also role for non-state actors	Diffusion of power among stakeholders
Rules	Sovereign states are the locus of authority and legitimacy	Sovereign states are the locus of formal authority and legitimacy; also informal authority and legitimacy for non-state actors	Stakeholders are the locus of authority and legitimacy
Discourses	Confrontational, dissentious	Mainly confrontational, dissentious	Mainly integrative, consensual

Table 6.2: Three ideal-types of global environmental policy arrangements

If we reconsider the cases in sections 6.3 and 6.4 - biodiversity and climate - it is obvious that both transcend the characteristics of an interstatist arrangement, given the fact that IGOs and non-state actors play important roles. To some extent, both domains also transcend the characteristics of an intergovernmental arrangement, given the role of local groups and the integration of discourse. However, the biodiversity arrangement seems to exhibit more transnational characteristics than the climate one. Whereas the former puts more emphasis on global-local linkages, the input of local groups and the integration of concepts (e.g. the concept of 'sustainable use'), the latter presents more traditional conflicts among states (e.g. on 'targets and timetables'). We therefore conclude that the climate arena exhibits more characteristics of an intergovernmental arrangement, whereas the biodiversity arena exhibits more characteristics of a transnational arrangement. However, this conclusion should be put into perspective immediately, since the structure of an arrangement can vary with each policy topic. For example, the North-South clash on forests in the biodiversity arena has even provoked *interstatist* types of (non-)policy making, whereas debates on flexible instruments in the climate arena are quite *transnational* in nature. Hence, one should be careful in categorising policy practices in types of arrangements, as generally old and new styles of policy making are combined in numerous ways.

As far as the process of political modernisation at global level is concerned, one may wonder whether a shift from interstatist to intergovernmental and, subsequently, from intergovernmental to transnational arrangements can be distinguished in the recent history of global environmental policies. If this shift is interpreted in linear, evolutionary terms, the answer should be 'no'. There is no evidence whatsoever that policy arrangements were, for example, only interstatist in the 1950s and 1960s, intergovernmental in 1970s and 1980s, and transnational in the 1990s, as the above examples on biodiversity and climate change show. However, if the shift is interpreted in a circular, both diachronic and synchronic, way, then the answer will be 'yes'. Global policies were definitely *more* interstatist in nature in the 1960s and are *more* transnational in nature today, but the former has not been entirely replaced by the latter. Features of old and new policy making go hand in hand nowadays and innovative and traditional ways of 'doing politics' are combined. Hence, political modernisation is a multi-dimensional process, in which the broadening of policy coalitions, the diffusion of power, the pluralisation of rules, and the integration of discourse take place, but in 'dialogue' with more traditional institutions.

6.7 CONCLUSION

Political modernisation is in this book understood as a transformation process in the political domain of society. It implies organisational and substantive change in policy arrangements. In this chapter, political modernisation has been described in terms of the broadening of policy coalitions, the diffusion of power, the pluralisation of rules, and the integration of discourse in global environmental policy arrangements. These concepts of 'broadening', 'diffusion', 'pluralisation' and 'integration' were deduced from empirical analyses of biodiversity and climate change policies on the one hand and global environmental policies in general on the other. From these analyses, it became clear that (1) new players - such as local NGOs and so-called user groups - have become involved in these policies; (2) big nations - such as the USA - do not necessarily dominate policy outcomes any longer; (3) the rules of the game today are less strict on the participation of non-state and local actors; and (4) environmental policies are being interpreted differently in the 1990s, e.g. in terms of sustainability and market-oriented instruments. Given these developments, policy arrangements are more transnational in nature now, although interstatist and intergovernmental characteristics may still be dominant in many cases of global environmental policy making.

REFERENCES

Adams, W. (1991), *Sustainable Development and the 'Greening' of Development Theory*, Niccos seminar 'Development Theories in the Nineties, Nijmegen: KUN.

Archer, C. (1983), *International Organizations*, London: Allen & Unwin.

Arts, B. (1994a), Dauerhafte Entwicklung: eine begriffliche Abgrenzung, *Peripherie. Zeitschrift fur Politik und Okonomie in der Dritten Welt*, 14,(54), pp. 6-27.

Arts, B. (1994b), The Survival of the Fittest? The North-South power struggle in the formation of the Convention on Biological Diversity, F. Schuurman, eds., *Current Issues in Development Studies. Global Dimensions of Agency and Structure*, Saarbrücken: Dreitenbach Publ., pp. 309-363.

Arts, B. (1998), *The Political Influence of Global NGOs. Case Studies on the Climate and Biodiversity Conventions*, Utrecht: Van Arkel/International Books.

Arts, B. and E. Mulleneers (1995), Vijftig jaar Verenigde Naties, vijfentwintig jaar mondiaal milieubeleid, B. Bomert and W. Van Genugten, eds., *Vijftig Jaar Verenigde Naties: Grote Problemen, Bescheiden Middelen*, Nijmegen: Studiecentrum voor Vredesvraagstukken, Cahier 66, pp. 125-156.

Beck, U. (1997), Global Risk Politics, M. Jacobs, eds., *Greening the Millennium? The new politics of the environment,* Oxford: The Political Quarterly/Blackwell, pp. 18-33.

Bergesen, H. and G. Parmann, eds. (1993), *Green Globe Yearbook 1993. An independent publication on environment and development from the Fridtjof Nansen Institute, Norway,* Oxford: Oxford University Press.

Bodansky, D. (1993), The United Nations Framework Convention on Climate Change: A Commentary, *Yale Journal of International Law,* 18(451), pp. 453-559.

Boehmer-Christiansen, S. (1994), Scientific Uncertainty and Power Politics: The Framework Convention on Climate Change and the Role of Scientific Advice, B. Spector, G. Sjöstedt and I. Zartman, eds., *Negotiating International Regimes. Lessons Learned from the United Nations Conference on Environment and Development (UNCED),* London: Graham and Trotham/Martinus Nijhoff, pp.171-180.

Burhenne-Guilmin, F. and S. Casey-Lefkowitz (1992), The New Law of Biodiversity, G. Handl, eds., *Yearbook of International Environmental Law,* London: Graham and Trotman/Martinus Nijhof, pp.43-59.

Caldwell, L.K. (1990), *International Environmental Policy. Emergence and Dimensions,* Durham, London: Duke University Press.

Carroll, J.E., eds. (1988), *International Environmental Diplomacy,* Cambridge: Cambridge University Press.

Chatterjee, P. and M. Finger (1994), *The Earth Brokers. Power, Politics and World Development,* London: Routledge.

Clegg, S.R. (1989), *Frameworks of Power,* London: SAGE.

Collins, K. and D. Earshaw (1992), The Implementation and Enforcement of European Community Environment Legislation, *Environmental Politics,* 1(4), pp. 213-249.

Duijse, P., A. Nentjes and J. Krozer (1998), *Verhandelbare CO2-Emissierechten,* Den Haag: Instituut voor Toegepaste Milieu-Economie.

ECOSOC (1995) E/AC.70/1995/CRP.1, Working Document for the Second Session of the Open-Ended Working Group on the Review of Arrangements for Consultation with Non-Governmental Organizations.

ENB (1997a), Report of the Third Conference of the Parties to the Framework Convention on Climate Change, *Earth Negotiation Bullitin,* 12(76).

ENB (1997b), Summary of the 19th United Nations General Assembly Special Session to Review Implementation of Agenda 21, *Earth Negotiation Bullitin,* 5(88).

Feld, W. and R. Jordan (1983), *International Organizations. A Comparative Approach,* New York: Praeger.

French, H.F. (1994), Making Environmental Treaties Work, *Scientific American,* (December 1994), pp. 94-97.

Giddens, A. (1984), *The Constitution of Society. Outline of the Theory of Structuration*, Cambridge: Polity Press.

Glowka, L., F. Burhenne-Guilmin, et al. (1994), *A Guide to the Convention on Biological Diversity*, Gland: IUCN.

Guzzini, S. (1993), Structural power: the limits of neorealist power analysis, *International Organization*, 47(3, Summer 1993), pp. 443-478.

Haas, P.M. (1993), Epistemic Communities and the Dynamics of International Environmental Co-Operation, V. Rittberger, eds., *Regime Theory and International Relations*, Oxford: Clarendon Press, pp.168-201.

Hocking, B. and M. Smith (1990), *World Politics. An introduction to international relations*, New York: Harvester/Wheatsheaf.

Hoogenboom, B. (1998), *Mexico and the NAFTA Environment Debate. The Transnational Politics of Economic Integration*, Utrecht: Van Arkel/International Books.

Hurrell, A. and B. Kingsbury, eds. (1992), *The International Politics of the Environment. Actors, Interests, and Institutions*, Oxford: Clarendon Press.

IPCC (1990a), *Policymakers Summary of the Scientific Assessment of Climate Change. Report to IPCC from Working Group I*, Geneva.

IPCC (1990b), *Policymakers Summary of the Potential Impacts of Climate Change. Report from Working Group II to IPCC*, Geneva.

IPCC (1990c), *Policymakers Summary of the Formulation of Response Strategies. Report prepared for IPCC by Working Group III*, Geneva.

Jacobs, M., eds. (1997), *Greening the Millennium? The new politics of the environment*, Oxford: The Political Quarterly/Blackwell.

Jacobson, H. (1984), *Networks of Interdependence. International organizations and the global political system*, New York: Alfred A. Knopf.

Johnson, S. (1993), *The Earth Summit: The United Nations Conference on Environment and Development*, London: Graham and Trotman/Martinus Nijhof.

Juma, C. (1989), *The Gene Hunters. Biotechnology and the Scramble for Seeds*, London: Zed Books.

Kaelberer, M. (1992), *State Power and International Environmental Cooperation: A Comparison of Collaborative Efforts on Global Warming and Ozone Layer Depletion*, ISA Convention, Princeton University, Department of Politics, Atlanta.

Käkönen, J., eds. (1994), *Green Security or Militarized Environment*, Dartmouth: Aldershot.

Kaufmann, J. (1980), *United Nations Decision Making*, Alphen aan den Rijn: Sijthof and Noordhof.

Keohane, R.O. and J.S. Nye, eds. (1989), *Power and Interdependence. World Politics in Transition*, Glenview: Scott, Foresman and Company.

Koch, K., R. Soetenendorp and A. van Staden, eds. (1994), *Internationale Betrekkingen. Theorieën en Benaderingen*, Utrecht: Het Spectrum.

Lélé, S. (1991), Sustainable development: a critical review, *Word Development*, 19(6), pp. 607-621.

Liefferink, J.D., P.D. Lowe and A.P.J. Mol eds. (1993), *European Integration and Environmental Policy*, London: Belhaven Press.

Lieshout, R. (1993), *Anarchie en Hierarchie. Een theorie van internationale betrekkingen en buitenlandse politiek*, Bussum: Coutinho.

Lukes, S. (1974), *Power. A Radical View*, London: Macmillan.

McCormick, J. (1989), *The Global Environmental Movement*, London: Belhaven Press.

Miller, M.A.L. (1995), *The Third World in Global Environmental Politics*, Buckingham: Open University Press.

Mintzer, I.M. and J.A. Leonard, eds. (1994), *Negotiating Climate Change. The inside story of the Rio convention*, Cambridge: Cambridge University Press.

Mitchell, J. (1998), *Companies in a World of Conflict*, London: The Royal Institute of International Affairs.

Morgenthau, H.J. (1993), *Politics among Nations. The struggle for power and peace*, New York: McGraw-Hill.

Nelissen, N., J. Van der Straaten and L. Klinkers, eds. (1997), *Classics in Environmental Studies. An Overview of Classic Texts in Environmental Studies*, Utrecht: Van Arkel/International Books.

Opschoor, H. (1989), Het nieuwe denken in het Kremlin en milieuveiligheid, *Transaktie*, 18(4), pp. 347-356.

Paterson, M. (1996), IR theory: Neorealism, Neoinstitutionalism and the Climate Change Convention, J. Vogler and M. Imber, eds., *The Environment and International Relations*, London: Routledge, pp. 59-76.

Princen, T. and M. Finger (1994), *Environmental NGOs in world politics. Linking the local to the global*, London: Routledge.

Ray, L.R. (1987), *Global Politics*, Boston: Houghton Mifflin.

Redclift, M. (1987), *Sustainable Development: Exploring the Contradictions*, London: Methuen.

Reinalda, B. (1997), *Private in Form and Public in Purpose: (I)NGOs as Political Actors in World Politics*, ECPR 25th joint sessions of workshops, Bern, Switzerland.

Risse-Kappen, T., eds. (1995), *Bringing Transnational Relations Back In. Non-State Actors, Domestic Structures and International Institutions*, Cambridge: Cambridge University Press.

Rowlands, I. and M. Greene, eds. (1992), *Global Environmental Change and International Relations*, London: Macmillan.

Sabatier, P.A. (1987), Knowledge, Policy-Oriented Learning, and Policy Change, and advocacy coalition framework, *Knowledge: Creation, Diffusion, Utilization*, 8(4), pp. 649-692.
Sachs, I. (1980), *Stratégies de l'écodéveloppement*, Paris.
Sanchez, V. and C. Juma, eds. (1994), *Biodiplomacy. Genetic Resources and International Relations*, Nairobi: ACTS Press.
Schermers, H.G. (1985), *Inleiding tot het Internationale Institutionele Recht*, Deventer: Kluwer.
Schermers, H.G. (1985), *Internationaal Publiekrecht voor de Rechtspraktijk*, Deventer: Kluwer.
Schneider, S.H. (1989), The Changing Climate, *Scientific America*, (September 1989), pp. 38-47.
Strange, S. (1988), *States and Markets*, London: Pinter.
Thompson-Feraru, A. (1974), Transnational Political Interests and the Global Environment, *International Organization*, 28(1), pp. 31-60.
Tolba, M.K. and O.A. El-Kholy (1992), *The World Environment 1972-1992. Two decades of challenge*, London: Chapman and Hall.
UNEP (1994), *Convention on Biological Diversity. Text and Annexes*, Geneva.
Waltz, K.N. (1979), *Theory of International Politics*, New York: McCraw-Hill.
Waters, M. (1995), *Globalization*, London: Routledge.
Weiss, Thomas G., David P. Forsythe and Roger A. Coate (1994), *The United Nations and Changing World Politics*, Boulder: Westview.
Weiss, L. (1997), Globalization and the Myth of the Powerless State, *New Left Review*, 225, pp. 3-27.
Wendt, A.E. (1987), The agent-structure problem in international relation theory, *International Organization*, 41(3, Summer 1987), pp. 335-370.
Wendt, A. E. (1992), Anarchy is what states make of it: the social construction of power politics, *International Organization*, 46(2, Spring 1992), pp. 391-425.
Willets, P., eds. (1982), *Pressure Groups in the Global System*, London: Frances Pinter.
Willets, P. (1996), From Stockholm to Rio and beyond: the impact of the environmental movement on the United Nations consultative arrangements for NGOs, *Review of International Studies*, 22(1), pp. 57-80.
Woerdman, E. (1997), Internationale onderhandelingen over klimaatbeleid: emissiereductie, Joint Implementation en verhandelbare emissierechten, *Milieu*, 1997(5), pp. 218-227.
World Commission on Environment and Development (1987), *Our common future*, Oxford, Oxford University Press.
Young, O.R. and G. Osherenko (1993), Testing Theories of Regime Formation: Finding from a Large Collaborative Research Project, V. Rittberger, eds., *Regime Theory and International Relations*, Oxford: Clarendon Press, pp. 223-251.

7. Green Planning: From Sectoral to Integrative Planning Arrangements?

JOAN JANSSENS AND JAN VAN TATENHOVE

7.1 INTRODUCTION

The notion of green planning is used in various ways in international literature. An influential definition of green planning is 'plans developed, mainly in industrial countries, to address escalating environmental problems' (Dalal-Clayton, 1996). In this definition green planning and environmental planning are considered synonyms. More recently, however, the notion of green planning has begun to incorporate a range of initiatives, including a variety of plans and strategies concerned with broader issues of sustainable development. In his study Dalal-Clayton concludes that although green planning is introduced in eighteen countries it has been done in eighteen different ways. However, most green plans produced in the industrialised countries remain focused on environmental issues; 'very few (mainly those undertaken independently of governments) have yet attempted to balance environmental, social and economic concerns - a central requirement of moving towards sustainable development' (Dalal-Clayton, 1996: 3). Every country shows different policy arrangements and practices of policy making to realise sustainable development. Green planning therefore seems to be a generic term; it embraces several (institutionalised) policy domains, which have the physical environment as their planning objective. Its material object is the sustainable development of society in its broadest sense, referring not only to ecological sustainability, but also to socially desirable and economically viable ways of sustainability. To realise such sustainable development the thus far fragmented policy initiatives have to be integrated. The formal objective of green planning involves its institutional dimension, which is the ability of governmental and other agencies to interfere and realise sustainable development (compare Zonneveld, 1991).

Green planning covers a variety of initiatives, differing in scope, status and impact. In the Dutch context it particularly implies *integrative* planning, which refers to initiatives to integrate the plans of different policy domains on the one hand, and to types of collaborative planning at regional level on the other. In this chapter we will focus on those initiatives in green planning which can lead to or will impede the institutionalisation of sustainable development in the Netherlands. In Dutch debates on planning the discourses on integration and on governance especially played a major role. In general, the discourse on integration refers to the integration of policy domains which have the physical environment as their policy and planning objective, such as spatial planning, environmental policy and water management, but more recently also nature conservation and some social and economic issues. More specifically, this discourse has also affected discussions about co-ordination and integration within and between the first three policy domains. The discourse on governance refers to the exercise of power and the use of rules to direct, control and regulate the activities of public and private actors. Governance is a dynamic concept. Changing relations between state, civil society and market influence its appearance, which is reflected in specific types of models of governing.

The central question in this chapter is: what has been the impact of the discourses on integration and on governance on the institutionalisation of green or integrative planning arrangements in the Netherlands? To answer this question we firstly have to understand the nature of the traditional planning arrangements within each of the separate policy domains of spatial planning, environmental policy and water management, in terms of coalitions, rules of the game and resources. Therefore, in section 7.2, a short history of the institutionalisation of these 'building blocks of green planning' is presented. Section 7.3 outlines the discourses on integration and on governance. In the Netherlands discussions on the integration of policy domains are not only influenced by but have also taken place within a context of changing discourses and practices on governance. The tension between integration of policy and new ways of governance is reflected in the types of arrangements in the policy domains of spatial planning, environmental policy and water management, and especially in the efforts to integrate these domains (section 7.4). In section 7.5 we present a twofold analysis, in which we discuss the possibilities of integrative planning as an instrument and as collaborative planning. The latter refers to experiments with new ways of governance in integrated (regional) projects. This analysis results in section 7.6 in an assessment of whether integrative planning will evolve into a suitable arrangement for an integrated environmental policy.

7.2. A SHORT HISTORY OF TRADITIONAL ARRANGEMENTS IN SPATIAL PLANNING, ENVIRONMENTAL POLICY AND WATER MANAGEMENT

7.2.1 Developments in physical planning

From its beginning in the 'Housing Act' of 1901 (see for example Mastop, 1984) untill the 1970s, the development of spatial planning in the Netherlands is highly representative of early modernisation. It was aimed originally at improving the poor quality of the urban environment produced by the unregulated rapid growth of cities at the end of the nineteenth century. Over the years spatial planning evolved into an instrument for state-initiated urban planning, controlled suburbanisation and the preservation of rural areas. The first goal was predominant from its origins until after the post Second World War, while the other two emerged between the 1960s and 1990s. Whatever its changing goals, spatial planning was intended to guide societal processes with the instrument of state-initiated plans. Strictly defined spatial planning 'refers to the taking of measures directly to influence the spatial disposition, and to the making of plans and other policy statements to guide these measures' (European Communities, 1999: 29). Plans were seen as desirable images and reflections of societal developments which had to be controlled, but also as crucial instruments to actualise these images and reflections (compare Van der Cammen and De Klerk, 1993). The National Plan (1944), the Working Commission for the Western Netherlands (1958) and the First and Second Policy Documents on Spatial Planning (1960 and 1966) were examples of the desirable directions in which society should develop. These documents provided a framework for societal developments during the period 1945-70, not only with regard to housing and urban development, but also in respect of the equitable and efficient distribution of spatial, social and economic functions over the country.

Planning doctrines and blueprint planning dominated physical planning in these periods (Faludi, 1980), during which the discourse of the 'manageable society' revealed itself in planning doctrines and principles for urban and rural areas (Korthals Altes, 1995; WRR, 1998). For the last four decades, physical planning in the Netherlands has been characterised by five basic principles. These are concentration of urbanisation (the location of urban activities), spatial coherence (keeping together interrelated functions and activities), spatial differentiation (to resist developments of uniformity, monotony and mono-functionality), hierarchy (to resist processes of levelling down, especially suburbanisation) and spatial justice (to strengthen the power of regions) (WRR,

1998). Despite the great variety of spatial concepts for urban and rural areas, they were all based on these fundamental principles. 'A spatial concept expresses through words and images in a summarized way, the view of a planning subject in respect of the desired spatial development of society and the nature of the interventions which are considered necessary' (Zonneveld, 1991a: 222).

During the 1950s and the 1960s the central planning doctrine was concerned with further orderly and regular urban growth. The impetus for central government to develop a planning strategy was threefold. First, the necessity of reconstruction of housing, industrial areas and infrastructure in the aftermath of the Second World War. Second, the migration from the agricultural areas in the north and the south which led to increased population pressures in the western parts of the country. Thirdly, the need to achieve some kind of co-ordination of the investment programmes of influential ministries such as Economic Affairs, Agriculture, Transport and Public Works, as well as the urban planning activities of the major cities.

Central strategic planning was supposed to play an important role, especially in preventing congestion in the Western Netherlands. The main features of Dutch strategic overall planning were the spatial separation of cities and city regions by green buffer zones, the preservation of the central agricultural area as a large-scale open space, and therefore the expansion of urbanisation in an outward direction. In this respect a clear distinction was made between the congested Western Netherlands and the rest of the country. This was translated into strategic spatial concepts such as the *Randstad*, the 'Green Heart' and 'rural areas'. The Western Netherlands was perceived as a *Randstad*, or 'rim city', with an open area in the middle, subsequently called the Green Heart. In contrast with metropolitan developments in countries such as Great Britain and Germany, in the Netherlands the accent was on 'district building' and controlled suburbanisation. However, the policy of controlling suburban growth was not successful. Its core idea, 'the principle of bundled deconcentration', based on grouping future developments within urban zones, was not realised.

Until the 1950s, spatial planning for rural areas focused on concentrating houses within the built-up areas of villages and towns on the one hand and nature conservation and 'the transformation of citizens' on the other. The notion of transforming citizens refers to a pedagogic and normative goal of prewar spatial planning. In the 1920s and the 1930s it was supposed that 'recreation in nature' was not only sensible from a public health perspective, but also provided a remedy for urban crimes and societal disturbances. The central assumption in spatial planning for rural areas was that they were threatened from the outside: by the encroachement of built-up areas. This focus changed during the 1950s, when the agricultural sector quickly modernised. Mechanisation, rationalisation of production and scaling-up put pressure on the ideal of a harmonious integration of agriculture with other rural functions such as nature, landscape and recreation. The growing dominance of agricultural production made it necessary to reorganise rural areas in such a way that modern techniques could be used. This resulted in the Land Consolidation Act of 1954, which made it possible to combine parcelling, opening up and the regulation of water. Land consolidation even evolved into an instrument for implementing complete regional development projects. The spatial organisation of rural areas so as to optimise agricultural production is reflected in the First and Second Policy Documents on Spatial Planning (respectively 1960 and 1966). In the Second Policy Document the emphasis was on the separation of agriculture and nature. Two kinds of rural areas were distinguished: agricultural production areas and areas with a mixture of agricultural production and landscape and nature conservation.

Population growth and the increase in industrial and agricultural production during the 1960s and the 1970s resulted in new problems for urban and rural areas, such as uncontrolled urbanisation, rush hour traffic and environmental pollution. At the same time nature conservation and environmental degradation emerged as issues on both the societal and the political agenda. Anti-modern critics forced spatial planning into a re-orientation of policy. For urban areas this was reflected in the notion 'the threat of suburbanisation' and for rural areas in 'the threat to nature and landscape'. As a result environmental issues were introduced in the Second Policy Document on Spatial Planning (1966). In addition to its former planning tasks the state now also had to provide care for and maintenance of a good quality of life, including a healthy environment.

In these decades the spatial planning system had been institutionalised in the Spatial Planning Act (WRO 1965). This Act, setting out procedures and jurisdiction for plans, can be seen as the first modern instrumental planning act (Drupsteen, 1982). It regulated the institutionalisation of spatial planning on the three administrative levels: state, provinces and municipalities (Van Schendelen, 1997). Within the planning system the local land use plans are crucial, while congruity with other policy domains and administrative levels has to be achieved through negotiations and co-ordinating mechanisms.

In the 1970s the quality of living conditions (liveability), individual emancipation and the emancipation of underprivileged groups became important goals and discourses in spatial planning. Spatial planning, as an example of facet planning[1], claimed the co-ordination of the diversity of sectors, such as

[1] The De Wolff Commission (1970) introduced the distinction between sectoral and facet planning. According to the Commission sectoral planning refers to public service provision and

agriculture, housing, nature and recreation. These new goals and discourses of spatial planning were elaborated upon in the Third and Fourth Policy Documents (1977 and 1988), a new Spatial Planning Act (WRO, 1985) and the introduction of the National Spatial Planning Key Decision (PKB), legalised in 1986 (see also section 7.4). As a result of the anti-modernisation claims of the 1970s spatial planning became an important political issue, while the demands of many pressure groups for participation in policy-making processes resulted in new ways of governance. On the other hand, attempts to strengthen the co-ordinating role of spatial planning arose from the discourse on integration, more specifically the failure of the sectoral organisation of policy and of earlier co-ordination mechanisms. In section 7.4 we discuss the institutionalisation of policy arrangements as a result of this interplay between integration and governance in the context of processes of late political modernisation.

7.2.2 Environmental policy in the 1970s: a lack of planning

In contrast with the policy domain of spatial planning, environmental policy in the 1970s had not yet developed a planning system of its own. During this first phase of institutionalisation the emphasis was on a process of structuration of the contents and the organisation of environmental policy (see chapter 1).

As in other European countries, the first environmental policy initiatives in the Netherlands date back to the nineteenth century and even earlier, with nationally varying environmental regulations *avant la lettre*, mainly focusing on industrial hazards and their impact on the neighbouring environment. We can point here to the Factory Act of 1875 - which became the Nuisance Act (1896) after it was combined with the Safety Act. Despite these early initiatives, the starting point of the institutionalisation of modern environmental policy was in the 1960s. This was when the structuration of both content and organisation of environmental policy could develop simultaneously as a result of three interrelated factors. First, the seriousness and magnitude of environmental problems increased enormously because of the rapid industrialisation after the

---

such like within the responsibility of individual ministers. Sectoral planning thus could be best organised as a departmental activity. Facet planning on the other hand refers to aspects, which the different sectors have in common, such as finance and space. Because sectors such as housing, agriculture, nature conservation, outdoor recreation, infrastructure and transport all lay claim to the common resource space, these claims have to be co-ordinated. According to the Commission spatial planning was a clear example of facet planning. The normative element of the sector-facet scheme was that facet planning was supposed to be superior to sectoral planning. Because the Spatial Planning Act at that time was the only legal instrument for this type of facet planning, spatial planners took the lead in governmental planning as such.

Second World War. Secondly, this economic growth was accompanied by an anti-modern cultural critique. Concerned scientists and new social movements attacked the capitalist system, its growth model and its destructive nature as regards the environment. Thirdly, environmental problems were no longer formulated as local issues only, in the way they were addressed in the early environmental legislation such as the Nuisance Act.

However, the early stage of the institutionalisation of environmental policy was largely dominated by the framing of environmental problems mainly in terms of public health issues. It was this threat to the public health that forced the state to act quickly. The collective feeling that 'something has to be done now', incited by scientists and environmental activists, was translated by the Dutch government into the Priority Memorandum on the Environment (1972), a policy document which formulated thirteen urgencies or priorities, concerning air, water and soil pollution. This 'urge to act' resulted in a sectoral approach to the areas of water, air, soil and organisms, thus reducing the content of environmental policy to sectoral legislation and the development and application of end-of-pipe technologies. The outcome of this was a long series of acts, each elaborated into specific norms and standards, and into a system of individual permits. Environmental policy was thought of as environmental legislation. Environmental planning was not an issue at that time.

The close relationship between environmental problems and public health issues was, as far as the Netherlands was concerned, also reflected in its organisational structure. In 1971 the Ministry of Public Health and Environmental Protection was formed, which included a Department for Environmental Management (DGM). This reinforced already existing organisational relationships between environmental and public health issues, for example environmental research being carried out by the national Institute for Public Health from the 1950s onwards and the Environmental Inspectorate (founded in 1962).

Because environmental problems were widely acknowledged as public health risks at that time, the institutionalisation of environmental policy as part of the health department seems quite 'natural' now. One should realise, though, that public health was not the only policy domain the early environmental concern addressed. In the 1970s a series of discussions took place on the position and co-ordination of environmental policy within spatial planning, public health, nature conservation and energy policy. At the same time the environmental movement suggested the creation of an independent Ministry of the Environment. The prominence of these interdepartmental struggles about the position and scope of environmental policy not only resulted in a fragmented institutionalisation of environmental policy in general, but it also hampered the development of an integrated environmental planning system. In short, these interdepartmental struggles did not lead to a clear decision on the position of environmental policy and its co-ordinating ambitions as a newly emerging 'facet' policy.

7.2.3 Developments in Water management

The battle against water is an integral part of the history of the Netherlands. Without dykes and polders the Netherlands as it is today would not exist. The control and management of water not only shaped much of the Dutch landscape, but also led to the institutionalisation of water management organisations. The District Water Boards, founded in the late Middle Ages, are still responsible for the quantitative and qualitative aspects of water management at regional level. More recent examples of the importance of water management can be found in the Public Works and Water Management Act of 1891 (resulting in the present National Agency for Transport and Public Works, better known as *Rijkswaterstaat*), the Public Works Act (1900) and the Rivers Act (1908).

Until the 1960s water management was largely dominated by quantitative aspects, including the prevention of floods (Jong, 1987). People's safety was the main concern, leading to huge technical projects, such as dykes and land reclamation schemes. As a result of the economic growth in the postwar period, paralleled by the emergence of environmental concerns, qualitative aspects e.g. water pollution became a central issue. In the First National Water Plan (1968) three key issues were identified: water protection, public water supply and agriculture. In this plan, which gave expression to the concern about the negative consequences of economic and population growth, several policy changes were suggested. One of these was to replace the inadequate legislation, still largely dating from the nineteenth century, which was mainly restricted to water quantity control. During the 1970s the planning of water quality management was assigned to the Ministry of Public Health and Environmental Protection, whereas the control of water quantity remained the responsibility of the traditional institutions of public (water) works. Water quality standards were set out by the Ministry in the Indicative Water Programme 1980-1984 (IPW, 1980). The Ministry of Transport, Public Works and Water Management published a second policy document, *Water Organisation*, in 1984. While this document acknowledges the need for an integrated water policy, for reasons of the interdepartmental division of tasks and competencies, it focuses on water quantity. Integrated policy is given further attention when the document addresses the need for a coherent policy on surface and ground water (Wisserhof and Crijns, 1992).

7.2.4 Traditional policy arrangements in the three policy domains

From the above it will be clear that the policy domains of spatial planning, environmental policy and water management have been institutionalised in very different ways (see Van Tatenhove and Van den Aarsen, 1996; Van der Vlist, 1998). Both spatial planning and modern water management date from around the beginning of the twentieth century. Until the 1960s the institutionalisation of these policy domains was strongly influenced by the characteristics of early political modernisation, as revealed in statist planning arrangements. Both plans and policy concepts could be seen as state-defined discourses, rules and resources, intended to guide societal and political developments. The future direction of these developments was incorporated in discourses and guiding principles such as safety and controlled suburbanisation. Managing the future was also characteristic of environmental policy, when this domain emerged at the beginning of the 1970s. However, the institutionalisation of environmental policy was from the start influenced by processes of both early and anti- political modernisation. While regulation and technical solutions were considered to be effective in overcoming environmental problems, these early political modernisation schemes were supplemented by anti-modern discourses. The more radical parts of the environmental movement especially fought for a new social order based on ecological and socialist principles. As a result the early development of environmental policy was affected by tension between social criticism and technological solutions, reflected in fundamental discussions about the scope of environmental politics on the one hand, and strains concerning its position in relation to other policy domains on the other. While spatial planning and water management already had developed mature, more detailed and more robust planning systems at that time, it was still crucial for the domain of environmental policy to define, delineate and legitimate its tasks, competencies and responsibilities.

Despite their different historical roots and characteristics, these policy domains were until the late 1970s largely dominated by statist arrangements, including in their planning systems. In the case of spatial planning, however, this picture has to be qualified. Whereas spatial planning at national level in the 1960s was still in the heyday of long-term blueprint or end-state planning, spatial planning at provincial level felt freed of its legal restrictions and embraced the new task of provincial structural planning as laid down in the 1965 Spatial Planning Act. In this Act the provinces were seen as the suitable level for the development of integrated spatial planning, serving as a vehicle for more general socio-economic development planning on a regional scale (Mastop, 1998). In general, however, the overall picture was that public actors, at different administrative levels, defined the central discourses and the rules of policy making, and they were supposed to be able to mobilise resources such as knowledge, manpower and money. In the background of all this was the optimistic belief in planning and regulation as a means of influencing political and social processes. At this high point of rational planning, all aspects of landuse, environmental pollution and water quantity were supposed to be manageable through regulation and, in the cases of spatial planning and water management, through all kinds of strategic, tactical and operational plans.

7.3 DISCOURSES ON INTEGRATION AND ON GOVERNANCE

At the end of the 1970s the deficiencies of the traditional statist arrangements became manifest. These problems concerned the implementation deficit, linked with a merely state-initiated planning system on the one hand and the fragmentation of policies, due to a lack of co-ordination and co-operation between different policy domains, on the other (see also chapter 1).

To overcome the deficiencies caused by the policy fragmentation, attempts were made to strengthen the co-ordination between the policy domains of spatial planning, environmental planning and water management, especially because politicians became more and more aware of the similarity in planning objectives of these policy domains. The main focus of this internal attempt at policy integration was resisting the differentiation of tasks and overcoming the struggle over the co-ordination of policies on the physical environment between a great variety of departmental organisations and divisions. In the next section we discuss the specific characteristics of the discourse on integration in the Netherlands. In this debate the emphasis is on the integration of policy aspects and its direction.

The implementation of this rather technical discourse in policy arrangements, however, was influenced by the discourse on governance. During the 1980s the relationships between state, civil society and market changed fundamentally. As a result of macro-processes, such as the economic crisis in the late 1970s and processes of globalisation in the 1980s and 1990s, the idea that the state could blueprint societal developments gradually disappeared. Due to these external factors the reflection on statist governance resulted in an erosion of the belief in detailed regulations, leading to political discussions about the scope and the level of detail and complexity of governmental regulation. But there was also an internal drive to change the statist planning arrangements and state-dominated ways of governance. The 'implementation deficit' and the uncertainties and unintended consequences of blueprint planning made clear that effective governance and governing implied more than state regulation and planning alone. To improve the implementation of plans and policies, interactions and new ways of co-governance with civil society and market were explored.

7.3.1. The discourse on integration

The last two decades have shown a dynamic development of initiatives for co-ordinating policies on the physical environment. The development of the environmental planning system during the 1980s, alongside the spatial planning system and the system of water management planning, confronted the national and provincial tiers of government with new problems of co-ordination. This was not just because the ensuing plans were complementary and to some extent competitive, but also because each of these policy domains had its own style of planning, procedures, types of regulation and planning concepts.

The tension between differentiation and integration, in particular, influenced the initiatives to co-ordinate policy and planning in the three policy domains. To understand the development of planning in the three domains, we will discuss the integration discourse as being the tension between differentiation and integration. Integration is the process in which separate parts are united, while differentiation refers to the splitting up into separate parts, which presupposes ways of specialisation in order to solve problems. In actual fact, integration and differentiation are two sides of the same coin, which are inextricably linked to policy and plans.

Analytically we can distinguish three elements in the integration discourse: the integration of policy *aspects*, the *direction* of integration and *stages* of integration. Integration of policy aspects concerns the integration of contents of policy, policy instruments, and planning and organisational elements. Secondly, integration can occur in different directions - internal, external, vertical and horizontal. *Internal* integration refers to the formulation of politics in which several sections, such as water, air and soil, within a policy domain are integrated. *External* integration applies to the co-ordination and integration of a policy domain with other domains. *Horizontal* integration is integration at the same administrative level, while *vertical* integration stands for integration between the administrative levels of national government, provinces and municipalities. Thirdly, a distinction can be made between different stages of integration. Stages of integration range from differentiation to integration. The following stages, showing increasing coherence, can be distinguished: differentiation, co-ordination, co-operation and integration. In the differentiation stage there is no coherence and the sectors are fully independent. During co-ordination procedures and administrative instruments achieve coherence, while the sectors remain largely independent. Co-operation is 'co-ordination plus': sectors are working together to formulate a partly mutual policy. In the final stage a new unity is created. No distinction can be made between the different sectors, which have merged.

In the initiatives to realise integrated green planning in the Netherlands each of the above mentioned stages can be distinguished, with a differentiated horizontal and vertical pattern, and even diagonal co-ordination and integration. Diagonal co-ordination refers to co-ordination procedures which cut through existing statutory systems, in order to ensure close co-operation between departments, tiers of government and if necessary private partners in the planning and realisation of complex and urgent strategic projects.

During the last three decades the integration discourse has forced the policy domains of spatial planning, environmental policy and water management to co-ordinate their planning activities more and more. Historically, spatial planning, as a facet policy, was supposed to be the leading policy domain for the integration of the various sectors. However, because in the early 1990s the water management and environmental planning systems also evolved into forms of facet policy, the sector-facet planning system was eroded. As we will discuss in sections 7.4 and 7.5 this resulted in the development of project-led planning, based on the principle of diagonal co-ordination.

7.3.2 New ways of governance

Solutions for the shortcomings of rational synoptic policy making were greatly influenced by discussions on state regulation in the 1970s and 1980s. In general, governance developed into a combination of forms of statist, reflexive and postmodern governing, in which interactions with the market and civil society were renewed and redeveloped.

Both anti-modernist and late or reflexive modernist discourses emphasised the need for participatory policy making, although in very different ways. While the radical democratisation perspective of the anti-modernists was aimed at the broadening of access to policy-making processes, in the late modernist perspective the rules of the game of policy making are also called into question. Gibbins and Reimer (1999) concisely formulate this change of perspective with regard to the position of social movements. According to them, the significance of social movements did not lie in what they accomplished, or in their opening up access to democratic politics to new individuals and groups, but, 'in their *style* of politics; a style that embraces performativity and politicization. By performativity is meant a set of rhetorical practices that encourage open dialogue, discussion, dissension and the sharing of information. By politicization is meant the process by which voices may bring previously excluded issues into politics' (Gibbins and Reimer, 1999: 113). On the one hand, parts of civil society, such as citizens and social movements, became more and more involved with public policy, decision making and planning; rules of the game shifted from goal-aim rationality towards political rationality, in which private actors had the opportunity to put forward their claims of interest in processes of negotiation and consensus making. On the other hand, the discourse 'more market, less government' was translated into a 'no-nonsense policy', which stressed decreasing public expenditure, privatisation, decentralisation and deregulation.

In general, the anti-modern critics of the 1970s, the ideologies of Thatcherism and Reaganism in the 1980s and the postmodern critics of the 1990s (see Gibbins and Reimer, 1999) changed the relationship between state, civil society and market, while new forms of governance were introduced. These late modern political and societal transformations led to a shift from state-dominated policy making towards ways of policy making in which market relationships and the participation of actors from civil society became more and more prevalent. The 'negotiating government', the 'calculating government' and the 'transactional government' became the dominant governance discourses, in which the emphasis was on ongoing interactions, negotiations and consensus building between public and private actors. Pestman and Van Tatenhove (1998) show that there is an increase in policy experiments based on reflexive governance. Reflexive governance is a type of rule-altering politics (see chapter 2), in which public and private actors have the ability to change the rules of the game and to mobilise resources in processes of policy making, in order to alter the foundations of statist arrangements based on rule-directed politics.

7.3.3 The dialectical relationship between integration and governance

Debates on integration and on governance have not taken place in isolation. The dialectical relationship between integration and new ways of governance has played an important role in the design and the implementation of innovative policy arrangements in the domains of spatial planning, environmental policy and water management. However, we have to make a distinction between internal processes of renewal of the politics of each policy domain on the one hand and the experiments to integrate aspects of these policy domains on the other. Both are deeply affected by the discourse on governance.

In section 7.4 we discuss the renewal of politics and the development of new planning and policy arrangements in spatial planning, environmental policy and water management in the 1980s and 1990s. This renewal of politics within each of the policy domains was, partly aimed at overcoming the deficiencies of the traditional statist arrangements. The introduction of liberal and plural policy arrangements, based upon reflexive types of governance such as network management and co-production, meant the strengthening of state, market and civil society relationships in policy making. For each policy domain we will analyse in which way the discourse on governance affected the statist arrangements. Were these traditional arrangements replaced by more plural and liberal arrangements? Or did this discourse result in a plurality of traditional and new planning arrangements?

During the 1980s each policy domain also had its own interpretation of integration, depending on its planning system and level of internal institutionalisation. The co-ordination and integration of policy on the physical environment became an important issue on the political and administrative agendas, forcing the adjacent policy domains to co-operate so as to achieve a common interpretation of sustainable development. In section 5 we discuss the two strategies which were used to realise an integrative policy for the physical environment. The first focused on the integration of planning systems and plans, especially at the provincial administrative level. The second implied the introduction of extra-legal policy experiments. Central to these experiments are the voluntary co-operation and co-ordination between public and private actors and the integration of policy issues from different policy domains. Examples are the experimental projects 'city and environment', 'integrated region-oriented policy', 'the water-system approach', 'sustainable building', and the like. By discussing these developments it is possible to answer the question of integrative planning arrangements will become institutionalised, or whether integrative green planning is no more than the integration of some aspects of the three policy domains.

7.4 NEW ARRANGEMENTS IN SPATIAL PLANNING, ENVIRONMENTAL POLICY AND WATER MANAGEMENT

7.4.1 Spatial Planning

In its search for solutions to the various planning problems caused by administrative centralism, the development of physical planning since the 1980s was very much influenced by political and societal developments. Socio-economic and demographic developments, such as a growing population, an expanding number of households, growing suburbanisation, unemployment, the reshaping of the welfare state and the increasing orientation on economic developments, were translated into new discourses and changing relations between state, civil society and market.

Firstly, there was a shift from the state to the market, or more precisely state-market relationships were being reinforced. Influenced by a predominantly economic discourse, liberal arrangements appeared in spatial planning in concepts such as the 'calculating' citizen, mutual obligations, public-private partnerships and the like. Secondly, the relationship between state and civil society changed. By introducing the concept of the 'open planning process' private actors were involved in early stages of the planning processes and were given the opportunity to change the rules of the game, mobilise resources and influence the context of planning.

The Fourth Policy Document on Spatial Planning (1988) is an example and a reflection of these changing discourses. In this document the Dutch government was looking for new opportunities resulting from international political and economic changes, especially the completion of the internal European market due in 1992. The core element was the contribution of spatial planning to economic growth. This was translated into policy aimed at improving the quality of the Netherlands as a country of transport and distribution, improving the quality of city regions, especially in the Randstad area, as a seat for international financial and business services and improving the quality of the Netherlands as 'a land of water'. The emphasis on transport and distribution was expressed in the upgrading of the quality of the main access points (the 'mainports' Schiphol Airport and the sea port of Rotterdam) and of the connections with other European countries (including the construction of high speed rail links with Belgium, France and Germany, as well as a new rail track between Rotterdam and Germany for freight transport) (European Communities, 1999). The mainport discourse (see also chapter 4) is further elaborated upon in the Fourth Policy Document-extra (1991), which deals with the implementation of spatial planning for the period 1995-2005. In particular, attention is paid to the development of new urban locations in order to reduce motor traffic, the development of infrastructure and the renewal of rural areas, while concepts such as 'the Netherlands as a land of water' and 'area-based integrated spatial and environmental policy' explicitly reflect the growing influence of the integration discourse.

The Fourth Document not only meant a change in policies and the ascent of plural and liberal arrangements, however, it also brought about a change in planning style. In the 1960s and 1970s, the co-ordination machinery of spatial planning was being constructed, and the principle of 'bilaterality' was institutionalised (Mastop, 1998). This machinery was based on vertical and horizontal co-ordination between all levels of government and for both facet and sectoral planning. Vertical co-ordination, laid down in the Spatial Planning Act, implied the development and application of national spatial policy in the spatial policies of provinces and municipalities, through formal and informal consultations. The instrument of the Spatial Planning Key Decision (1986), especially the national structure plan for a broad aspect of spatial policy (*structuurschets*) and the national structure plan for one policy sector (*structuurschema*), played an important role in the horizontal co-ordination of spatial planning. These plans institutionalised the principle of bilaterality, which refers to the equal importance of sector and facet policy.

The change in planning style generated by the Fourth Document involved the introduction of types of project-led planning based on the principle of diagonal co-ordination. What this means is 'co-ordination procedures that cut through existing statutory systems in order to ensure close co-operation between departments, tiers of government and if necessary private partners' (European Communities, 1999: 3). Diagonal co-ordination has been introduced for the planning and execution of complex and urgent strategic projects. Project-led planning is a new planning style, which does not rely on the procedures of the statutory system but is based on often extra-legal types of communicative and negotiation planning. The vital elements of this type of planning, such as network management, creating 'win-win' situations and public-private partnerships based on civil law covenants, meant a further strengthening of state, civil society and market relationships within the system of spatial planning. Infrastructure projects like the Betuweroute and high speed rail links are recent examples of project led-planning, as chapter 4 showed. Other examples are 'city and environment' projects and integrated region-oriented projects (see section 7.5).

Another example of the changing discourse in spatial planning and the implementation of open or interactive ways of planning is the project 'The Netherlands 2030' (VROM, 1997). In this national debate the likely spatial organisation of the Netherlands in 2030 was discussed by representatives of all levels of government and a variety of private actors. The debate focused on four scenarios. In each scenario different choices were made concerning the demands for space, infrastructure, environment and economics, ways of living and residential preferences, nature and landscape and the relationship nation-state and civil society.

In general, state-initiated planning and rule-directed negotiations were accomplished, with forms of integrative and interactive planning based on reflexive ways of policy making in the 1990s, while flexible and easily adjustable plans were introduced. Planning was more and more supposed to guide or to accompany social processes through flexible plans. Global outlines for future developments were translated into a diversity of projects and concepts. This change in planning had its effect on the planning system itself. Dutch spatial planning today is characterised by its decentralised nature and the co-ordination system. Co-ordination has to be achieved through negotiations, by the convincing of and co-operation between the administrative levels and private actors, following legislative procedures. In 1998 the Dutch Scientific Council for Government Policy (*Wetenschappelijke Raad voor het Regeringsbeleid* - WRR) suggested a new planning system, which it called spatial development politics. The starting point would be a dynamic approach to spatial planning, with the key concepts of differentiation and selectivity. General planning concepts, such as urban junctions and the compact city, pay insufficient attention to societal changes, and differentiation refers to the introduction of 'argumentative concepts'. These open concepts require a region-specific interpretation with its own argumentation, which will increase the possibilities of regional co-ordination by many stakeholders. On the other hand national planning should be more strategic and more selective. Central government should formulate the main spatial structure, consisting of three kinds of areas: basic areas, development areas and national projects (WRR, 1998).

7.4.2 Environmental Policy

In 1982, the Department Environmental Management was moved from the Ministry of Public Health and Environmental Protection to the Ministry of Housing, Spatial Planning and the Environment (VROM). This organisational stabilisation reinforced a process of (re-)structurering of the contents of environmental politics. In this process of structuration the emphasis was on the development of environmental planning, integration and new ways of governance.

In the documents *Integration Plan Environmental Policy* (PIM, 1983) and *More than the sum of its parts* (1984) an analysis was given of the fragmented institutionalisation of environmental policy and its problems. The overall conclusions were that there was a lack of long-term policy, there were too many plans with no coherence, and the relationships between plan and the other instruments were unclear. On the basis of this analysis a distinction was made between internal and external integration, in the *Integration Plan Environmental Policy*. The document *More than the sum of its parts* (1984) can be considered as the actual start of the environmental planning system. This document, a reaction to the fragmentation and diversity of sectoral legislation and plans, suggested a planning system consisting of an integrated strategic policy plan and an operational implementation programme at national, provincial and local levels.

At national level, three Indicative Environmental Programs (IMPs) had been published during the 1980s, followed in 1989 by the first National Environmental Policy Plan (NEPP). In this plan sustainable development is the main objective of environmental policy, worked out in detail in a series of actions.

During the 1980s the development of environmental policy and planning was strongly influenced by the discussions about internal and external integration. On the one hand there was the need to integrate the fragmented environmental legislation and to reduce problems of cross-media transfer (internal integration), while at the same time the relationship with other policy domains and private actors had to be clarified.

Illustrative examples of internal integration are the integration of legislation and the development of integrative policy concepts. The starting point for the integration of environmental legislation was the General Environmental Regulation Act of 1980, which was followed by the Environmental Management Act (1993). In this second Act integrative elements were added, such as regulations concerning finances, penalties, the planning system, and permits and general rules (Michiels, 1993). An evaluation of this Act demonstrated, however, that the process of internal integration of environmental policy had not yet been completed and that planning is not a panacea for achieving integration (Coenen and Janssens, 1997). To overcome problems of cross-media transfer new integrated concepts were developed. Separate sections, such as soil, water and air, were replaced by integral themes, such as acidification, squandering and depletion, while a region-oriented environmental policy was also developed, which involved a more explicit overall environmental policy in specific areas (groundwater protection, soil protection).

In addition to the central task in the 1980s of realising internal integration, external integrative initiatives were also taken. At national level this was done through new policy concepts, in which the relationship between state, civil society and market was intensified, and the responsibility of polluters to formulate their own environmental policies was strengthened. Examples of these policy concepts are the internalisation of responsibilities, the target groups policy and the integrated region-oriented policy. Target group policy and integrated region-oriented policy are, as well as an expression of external integration, also examples of new ways of governance in environmental policy (Van Tatenhove, 1999). Although the concept of internalisation was originally introduced to improve the implementation and enforcement of environmental legislation in the industrial sector, it soon evolved into a strategy to encourage polluters' environmental responsibility. Internalisation refers to the need to increase the willingness of polluters to contribute to the solution of environmental problems by invoking their sense of 'social responsibility', while at the same time paying serious attention to the needs and wishes of these actors. The concept of internalisation is further developed in the target group policy. Target groups are defined as more or less homogeneous groups of polluters, such as agriculture, traffic and transport, industry and refineries, gas and electricity supply. By approaching these target groups directly, an attempt is made to bring about an internalisation of environmental perspectives among them and to achieve environmental policy objectives. In the integrated region-oriented approach an integrated (environmental, spatial and water) policy is formulated at local levels through negotiations between public and private actors.

This shift from internal to external integration is however not without problems. For example, at provincial level co-ordination of the strategic plans (Regional Spatial Plan, Environmental Policy Plan and Water Management Plan) has remained inadequate. In theory co-ordination is accomplished through the so-called 'leapfrog principle': the next plan published should take into account the decisions and consequences of the previous one. In policy practice not only did the sectoral plans remain in existence, the shift towards external integration also had a negative impact on the internal integration of environmental policy. Bouwer (1997) stated that by emphasising external integration the failure of internal integration is deliberately covered up. The above-mentioned evaluation of the Environmental Management Act showed that not all environmental aspects could be integrated into one policy, because of the differences between them.

The 1990s was a period of consolidation and renewal of environmental policy. The further development of the target group and integrated region-oriented policies meant a consolidation of environmental policy, while at the same time quite basically altering its governance principles. The latter can be best summarised in the concepts *broadening* and *flexibility*. Broadening refers to the desire to achieve integral green planning, in which sustainability is an integrated element of other sectoral activities and policies. Flexibility means the development of interactive and reflexive types of policy making. More and more both formulation and implementation of environmental policy is the result of interactions and negotiations between public and private actors. Important events in this renewal of the content of environmental policy were four conferences organised between 1990 and 1996, at which representatives from the Ministry of the Environment, the Union of Water Boards, the Inter-provincial Consultative Body and the Association of Dutch Municipalities were present. The main conclusions of these conferences were that environmental policy is the concern of all involved and no longer a state monopoly alone, while under certain conditions deviating from environmental norms is acceptable. This new direction was detailed in the Second National Environmental Policy Plan (NEPP2, 1992) and more especially in the Third Plan (NEPP3, 1997). The NEPP2 still emphasised the implementation and further development of target group policy, because internal and external integration and implementation remained bottlenecks in environmental policy. In the NEPP3 emphasis was also given to stimulating sustainable patterns of production and consumption, differentiating environmental norms for different regions, and external integration with for example physical planning and water management. Not only did environmental policy making during these years change from direct regulation by legislation to interactive and reflexive forms of governance, new arrangements were also formed at regional and local levels (see section 7.5).

7.4.3 Water Management

The planning system in water management is based on the Water Management Act of 1989. Central to this Act is the intention to create coherent and effective policies, by introducing an extensive planning system and co-ordination procedures. The provinces were given a core position in water management, and were obliged to draw up strategic water management plans. However, the legislation and planning system provided inadequate means for realising horizontal (between the provincial policy arrangements) and vertical (between the administrative levels) co-ordination (Kuijpers and Glasbergen, 1990). During the 1980s and 1990s the emphasis in water management moved from internal to external integration. These changes were reflected in new planning concepts, such as integral water management and the water system approach.

Integral water management, introduced in the IPW 85-89 and further elaborated upon in the document 'Dealing with water' (1985), concerns the integrated management of water systems. The aim of this water system approach is to find coherence between water quantity and water quality policies on the one hand, and to develop an integrated approach to surface and ground water on the other. In this so-called functional coherence ecological aspects were emphasised, although the accent gradually changed from considering ecological aspects only (internal integration) to also taking other aspects, such as landuse planning and environmental policy (external integration), into account. Management and control of the water systems became the joint responsibility of a variety of actors, and new coalitions were formed in controlling and managing water systems (Van der Vlist, 1998, Havekes et al., 1990). The Third and Fourth Policy Documents on Water Management describe integrated water system management, based on the water system approach, as coherent water management (policy and control) by several administrative levels with strategic water management tasks. The underlying principle is that greater cohesion is required in water, environmental, nature and spatial planning policy, at both macro and micro level. Additional money is being set aside for the cleaning up of polluted riverbeds and for nature development. A total of 700 million guilders has been made available up to 2010 for the development of natural areas in the Rhine-Meuse delta and the IJsselmeer, and for restoration of and planning measures for other rivers and lakes.

More coherence between landuse, environmental and water management policy in particular is considered necessary to achieve the policy goals. Nowadays, water is not only regarded as a regulative principle in spatial planning, but also as one of the frameworks for the integration of several policy domains, such as environmental policy, nature conservation, agriculture and the like. An expression of these changes is the region-oriented policy approach, which is translated into policy concepts such as '(spatial) planning with streams', 'water in the city' and 'blue junctions'. Blue junctions are nodes in a water system, where water flows from one basin to another, for which reduction targets are formulated. An example of an approach based on the 'blue junctions' concept is the design and development of an integrated vision for the Regge river basin in the Netherlands (Van der Vlist, 1998). In this vision spatial hydrological zoning is combined with reflexive ways of governance. Because the region is defined as a bubble, differentiated norms can be formulated for each node to solve the problems of eutrophication and the fall in the hydraulic head. In this planning practice the participation of private actors is important, in order to achieve a jointly-accepted vision and the implementation of this vision in the everyday practices of private actors.

7.5 INTEGRATIVE PLANNING ARRANGEMENTS IN THE NETHERLANDS

7.5.1 Provinces and the integration of plans

One aspect of green planning concerns initiatives to integrate plans from different policy domains. The provinces are expected to play a key role, because this administrative level is seen as the most appropriate governmental tier to be central in the integration of physical planning, environmental policy and water management. There are two main reasons for allotting this role to the provinces. One is their involvement with all sectoral plans (in each of which they have to translate national goals into operational programmes). The other is their traditional role of linking local and national policies. In addition, the provinces also have an important part in the co-ordination between the administrative levels. In this respect, provincial planning is used to improve the position of the provincial administrative level and the implementation of new tasks, such as legitimacy through counselling, integration through directing and partnership.

From the 1990s onwards the integration of planning and policy was supposed to be a provincial and local challenge. The provincial administrative level developed initiatives to unite the regional spatial, provincial water management and provincial environmental plans into one integrated plan for the physical environment. It is striking that in recent discussions other policy domains, such as infrastructure, nature conservation and economic policy are also involved. In a few provinces social and welfare politics have even entered the policy arena. However, the institutionalisation of integrative green planning in the provinces varied from the development of integrated plans, via the drawing up of 'umbrella' documents, to a strengthening of the leapfrog principle. An umbrella document can be seen as a coherent integrated guide for future spatial, environmental and water planning (Janssens and Weisz, 1998), whereas in integrated plans the sectors have been integrated. An important function of these more or less integrated plans is to formulate meta-strategic policy targets. An example of such an integrated plan is provided by the province of Drenthe (1999).

Although the provincial level is the most promising level for co-ordination and integration of plans, provinces have to deal with a number of problems. First, the participating policy domains have different (conflicting) systems of legislation. Secondly, each policy domain is characterised by a specific political-administrative system and by its own policy culture, which are difficult to change. Because of these institutional and cultural aspects, the integration of plans and policies of the separate domains has resulted in several problems, from which juridical, instrumental, conceptual, organisational and political obstructions ensued (Janssens and Bouwer, 1996). Although there appear to be some fundamental juridical obstructions, it is generally understood that the legally foreseen degrees of freedom are sufficient to improve co-ordination and integration. Mastop (1997) concludes that the integration challenge has multiple dimensions, which can be found in several integration mechanisms. Examples are the planning systems, co-ordination and correction mechanisms, (communication) instruments, concepts, etc. The multiple dimensions of the integration challenge mean that all aspects (content, planning, instruments and organisation) have to be integrated.

There is a resemblance with the developments in the 1960s, when comprehensive planning was seen as the panacea. The planning lessons of those years seem to have been forgotten, as can also be seen from international planning literature. The first lesson is the impossibility of rational comprehensive planning. People are confronted with their bounded rationality. Planners cannot co-ordinate and understand all consequences, while the number of elements that are to be integrated has to be restricted. A second lesson is that integration can be realised only when it is really necessary. Thirdly, planners must avoid plans that have strategic as well as operational elements, as it is unwise to combine too many functions in one plan. The fourth important lesson is that the planning process is far more important than the planning document. During the planning process (often lasting two or three years) there is a vast amount of exchange and co-ordination between policy domains. In the planning process special attention has to be paid to the role of key persons in the formation and implementation of policy. They form an informal network and are important for the performance of the plan (Coenen and Janssens, 1999). The overall conclusion is that the opportunities for policy integration through planning, although important, are limited.

These lessons are important for the further development of the integration of plans at provincial level. Planning can play an integrative role, but cannot solve all problems. As we have seen, traditional policy arrangements are not able to cope with the new problems and societal developments. Social, economic and technological developments, such as technological innovations, information technology, demographic factors and the like, will have their impact on the institutionalisation of green planning arrangements. The new planning arrangements have to cope with the rapidly changing circumstances by introducing flexibility in the rules of the game and in the selection of coalitions. This multi-dimensioned integration challenge also has an effect on the implementation of organisational structures, resulting for example in the more frequent use of project organisations and structures.

7.5.2 Integrative planning experiments

At national level the call for integrative regional planning experiments coalesced around specific problems, such as the perceived hampering of economic and spatial developments by environmental standards, the enormous environmental pressure in some specific areas (e.g. Schiphol and Rotterdam) and the lack of space in others. To overcome co-ordination and co-operation problems in these areas a regionalisation of policy was developed, resulting in a diversity of integrative regional arrangements.

Integrative regional planning is currently flourishing in the Netherlands. Two types of such planning can be distinguished: regional projects with a predominantly regulatory nature and projects that are characterised by a more or less integrative approach and reflexive forms of governance. The first category comprises groundwater protection areas, nature monuments and wetlands, where the emphasis is on relatively strict rules for environmental protection. The second category consists of, for example, area-based integrated spatial and environmental policy (ROM) and a group of rather heterogeneous integrated region-oriented projects such as 'city and environment' and ROM-like projects including Valuable Man-made Landscapes (WCL) and Rural Area Pilot Plans (SGP).

The ROM-projects were initiated and implemented at the beginning of the 1990s. They represent a particular group of ten areas where great pressure on the environment and spatial qualities exists or is expected to occur in the near future. In these projects governmental agencies and private parties in the ROM-area are invited to draw up a working plan to solve acute environmental problems, improve spatial planning characteristics and implement a package of concrete measures and promising socio-economic projects. The official start of a ROM-project is marked by a covenant between all partners, while public and private actors negotiate to formulate the integral vision and the package of measures. In 1992 the Ministry of Agriculture, Nature Management and Fisheries introduced WCL-areas, to promote projects that would combine activities in farming, nature management, landscape protection, cultural heritage and tourism. Eleven regions were designated WCL-areas and asked to submit projects which would be subsidised between 1994 and 2001. Initially the farmers were rather suspicious about yet another 'special area' designation, as they feared it would involve additional constrains for them to cope with. However, the acceptance of subsidised nature management by farmers and 'broadening' agriculture by diversifying into the use of traditional farming methods, agro-tourism or an on-farm range of 'regional' products, gradually increased. SGP-projects were initially aimed at nature development. However, rural inhabitants opposed the large scale and the seeming one-sidedness of the proposed plans from ministries and provinces. In most cases the policy plans were modified, as a result of negotiations with private actors. As well as nature development, agriculture, recreation and sometimes housing are considered as issues for rural area pilot projects.

From an evaluation of ROM-policies it appears that production-oriented projects (aimed at the economic development of agricultural industry) fare significantly better than implementation projects intended to address nature and environmental qualities (Novioconsult, 1998). In both ROM- and ROM-like projects several objectives are supposedly integrated, thus combining environmental policy, spatial planning, water management and economic policy. Results are difficult to assess, as the aims of these regional development policies are mostly framed in terms of intention. It is therefore hardly surprising that the monitoring of region-oriented policies is greatly deficient (Driessen and

Groenenberg, 1998). The innovative quality and additional value of the region-oriented policies is considered to lie mainly in their specific mode of organisation and governance, marked by consensus-building, combining of different interests and proceeding along the lines of regional development projects (Frouws and Van Tatenhove, 1999).

The 'city and environment' projects are based on the paradox of the compact city, which means that the development of a city is both a blessing and a curse for its liveability (VROM, 1995). The aim of the projects is to improve the quality of live in the city and to use urban space in a more intensive and varied way. In this region-oriented approach the focus is on the environment and on spatial planning. An innovative aspect of the city and environment experiments is the concept of compensation. Central to this concept is the notion that environmental standards can be weighed against other standards. In many cities developments are hampered by the application of generic environmental norms. In the about twenty 'city and environment' experiments the participants were asked to show creativity and flexibility, firstly in trying to solve the problems within the scope of environmental standards and legislation, and secondly in finding a solution through compensation on certain conditions. The compensation has to be found in the first instance through other environmental measures, while in special cases it may be achieved through the provision of other societal benefits. Compensation for environmental standards is a disputed issue. Bouwer (1996) states that there are still too many uncertainties in terms of environmental efficiency. The effectivity and legitimacy of the compensation principle are also questionable. Who and what will be where and when compensated?

When taking an overall view of the integrative planning experiments, some promising results of this type of governance can be reported. These relate to the *formulation* of views and measures which do justice to regional circumstances and problems. Moreover, creative solutions have been invented, combining environmental improvement and the economic renewal of rural and urban areas. The negotiating over regional development projects has also contributed to the accommodation of conflicting interests. Initial plans, proposed by ministries and provinces, were thus adjusted. This occurred for example in rural area pilot projects, where the emphasis on nature development was moderated in favour of agriculture, recreation and housing (Goverde, Wisserhof et al., 1997). Introducing greater flexibility, for instance in existing legislation on noise and ammonia emissions, will offer better opportunities for urban and agricultural development. However, as with many of these projects implementation started only recently, we have the impression that good intentions, focusing on processes of negotiation and co-operation, prevail over substantial and sustainable development in the attempts to solve region-based environmental, economic and spatial problems (compare Frouws and Van Tatenhove, 1999).

7.6 CONCLUSIONS

The development of policy for the physical environment historically has been, and still is, the object of a number of domains. In this chapter we analysed the institutionalisation of spatial planning, environmental policy and water management and the initiatives within and between these policy domains to develop integrative policy and planning arrangements for the physical environment. The traditional arrangements in these domains proved to be inadequate to solve environmental, spatial and water management problems in an integrative way. This was partly because of the fragmented institutionalisation of the three planning systems and the barriers between them, partly because of the limits inherent in rational comprehensive planning.

The discourses on integration and on governance offered a way to develop integrative policy and planning arrangements. They opened a policy window, enabling actors to look for and take account of new directions in the content and the organisation of green planning. Not only were the many different aspects of the physical environment, such as land use, environmental nuisance, water quality, nature protection, and the connections between them (Needham and Janssens, 1998), taken into account, but these substantive initiatives were also translated into organisational experiments. During the 1990s a diversity of experiments, which had the integration of environmental policy, physical planning and water management as their main focus, was developed looking for new coalitions, new rules of the game and new resources. Firstly these initiatives concerned the integration of plans. Especially at provincial level, but also at local level, there are some promising results in attempts to integrate the plans of the three policy domains. However, the distribution of power between the actors in these newly formed coalitions is unequal. There is empirical evidence that spatial planning still dominates integration processes. Secondly there is great variety in integrative planning experiments. The experiments with ROM, WCL and 'city and environment' showed an increasing regionalisation of policy for the physical environment, in which public and private actors were offered the chance to negotiate on the rules of the game and to mobilise resources. Within the framework of national policy, this resulted in some cases in innovative definitions of problems and measurements for a specific area. These experiments, as well as the development of other instruments such as permit co-ordination, are perhaps the forerunners of an even more integrative approach to achieving sustainable development.

A central question in the years to come will be whether the experiments with integrative planning can offer solutions for the environmental, water and spatial problems. And will these integrative green planning arrangements be institutionalised? On the one hand history has taught us the limits of planning, and this could temper the expectations. On the other hand we see some promising developments, both at provincial level and within the extra-legal forms of integrative regional planning. However, we have to keep in mind that although the role of planning has changed over the years, the plan as a well-institutionalised instrument still exercises a strong influence over the rules of the game and in consequence over the formation of policy arrangements. In this institutional context promising experiments, both from the perspective of integration and of governance, always have to compete with existing legislative and planning frameworks. This reminds us of what Offe (1996) refers to as 'the state's ordered retreat', which requires not only a *strong* state which is able to establish and maintain negotiation procedures replacing the state's authority, but which also presumes that societal actors will accept public duties and formal competencies. The transfer of state competence to societal coalitions in the field of integrative green planning may bring about the formation of plural arrangements, in which regional legitimacy, regulatory responsibilities, the development of relevant expertise and administrative capacity are the responsibility of all actors involved in the realisation of sustainable development.

REFERENCES

Bouwer, K. (1996), *Beleidsmatige aspecten van compensatie in het stedelijk gebied*, Den Haag: NIROV.

Bouwer, K. (1997), *Van milieubeleid naar omgevingsbeleid?*, Nijmegen: KUN.

Coenen, F.H.J.M., J.R. Janssens and H-J.W. Oosterveld (1998), *Provinciale milieubeleidsplanning als integrale paraplu? Een evaluatie van de bereikte interne integratie in het provinciaal milieubeleid door het hoofdstuk Plannen van de Wet milieubeheer*, Den Haag: VROM, ECW-reeks, Achtergrondstudie nr.33.

Coenen, F.H.J.M. and J.R. Janssens (1999), De grenzen van integreren van beleid door planning: Ervaringen met provinciale milieubeleidsplanning, *Beleidswetenschap*, 13(2), pp. 107-124.

Dalal-Clayton, B. (1996), *Getting to grips with green plans. National-Level Experience in Industrial countries*, London: Earthscan Publications Limited.

Driessen, P.P.J. and M.C. Groenenberg (1998), *Monitoring van gebiedsgericht Milieubeleid. Een analyse op provinciaal niveau*, Utrecht/Bilthoven: RIVM.

Drupsteen, T.G. (1982), *Milieubeleidsplanning in relatie tot recht*, oratie, Zwolle.

European Communities (1999), *The EU compendium of spatial planning; systems and policies. The Netherlands*, Luxembourg: Office for Offical Publications of the European Communities.

Faludi, A. (1980), *Implementation planning or the implementation of plans? Effectiveness as a methodological problem*, Amsterdam: UvA/PDI.

Frouws, J. and J. Van Tatenhove (1999), *Regional Development and the Innovation of Governance*, Workinggroup 4, EU-Cost A-12, Vienna, Austria.

Gibbins, J.R. and B. Reimer (1999), *The Politics of Postmodernity. An Introduction to Contemporary Politics and Culture*, London: SAGE.

Goverde, H.J.M., J. Wisserhof, E.K. Dijkstra and R.A.M. Tilmans (1997), *Bestuurlijke evaluatie strategische groenprojecten natuurontwikkeling*, IBN-rapport 269, Wageningen: IBN-DLO.

Havekes, H.J.M. (1990), *De Wet op de Waterhuishouding: integraal waterbeheer in juridisch perspectief*, Lelystad: Kon. Vermande.

Janssens, J.R. and K. Bouwer (1996), *Naar één provinciaal omgevingsplan !!?, Een bijdrage aan het debat over omgevingsbeleid*, Nijmegen: KUN.

Janssens, J.R. and M. Weisz (1998), Provincie op weg naar beleid voor de fysieke omgeving, *Beleidsanalyse*, (27-2), pp. 4-11.

Jong, J.D. (1987), *Water and Land Management in the Netherlands: History, Present Day's Situation and Future. Water for the Future: Water Resources Developments in Perspectives*, Rotterdam: W.O.A.J.E.P., Wunderlich, Balkema.

Korthals Altes, W. (1995), *De Nederlandse planningsdoctrine in het fin de siècle. Voorbereiding en doorwerking van de Vierde nota over de Ruimtelijke Ordening (Extra)*, Assen: Van Gorcum.

Kuijpers, C.B.F. and P. Glasbergen (1990), *Perspectieven voor integraal waterbeheer*, Den Haag: SDU.

Mastop, J.M. (1984), *Besluitvorming, handelen en normeren: een methodologische studie naar aanleiding van het streekplanwerk*, Bunnik.

Mastop, H. and L. Van Damme, et al. (1997), *Instrumentatie van het restrictief beleid*, Nijmegen: KUN.

Mastop, H.J.M. (1998), *Dutch National Planning at the Turning Point. Re-thinking Institutional Arrangements*, International Seminar on National-level Planning Institutions and Decisions, Haifa, Israel.

Michiels, F.C.M.A. (1993), *De Wet Milieubeheer*, Zwolle: Tjeenk Willink.

Ministerie van Verkeer en Waterstaat (1985), *Omgaan met Water*, Den Haag: SDU.

Ministerie van Volkhuisvesting, Ruimtelijke Ordening en Milieubeheer (1988), *Vierde Nota Ruimtelijke Ordening deel A*, Den Haag: SDU.

Ministerie van Volkshuisvesting en Ruimtelijke Ordening (1960), *Eerste nota over de Ruimtelijke Ordening*, Den Haag: SDU.

Ministerie van Volkshuisvesting en Ruimtelijke Ordening (1966), *Tweede Nota over de Ruimtelijke Ordening*, Den Haag: SDU.

Ministerie van Volkshuisvesting en Ruimtelijke Ordening (1977), *Derde nota over de Ruimtelijke Ordening. Deel 3a: beleidsvoornemens over ontwikkeling, inrichting en beheer*, Den Haag: SDU.

Ministerie van Volkshuisvesting, Ruimtelijke Ordening en Milieubeheer (1990), *Vierde Nota Ruimtelijke Ordening Extra, deel 1*, Den Haag: SDU.

Ministerie van Volkshuisvesting, Ruimtelijke Ordening en Milieubeheer (1995), *Stad and Milieu, rapportage. Waar vele willen zijn, is ook een weg*, Den Haag: VROM.

Ministerie van Volkshuisvesting, Ruimtelijke Ordening en Milieubeheer (1997), *Nederland 2030. Discussienota; verkenning ruimtelijke perspectieven*, Den Haag: VROM.

Ministerie van Volksgezondheid en Milieuhygiëne (1980), *Indicatief Meerjarenprogramma Water, 1980-1984*, Den Haag: SDU.

Needham, D.B. and J.R. Janssens (1998), *Omgevingsbeleid: Een internationaal verschijnsel? Zes landen vergeleken*, Nijmegen: KUN.

NovioConsult (1998), *Eindrapportage evaluatie ROM-aanpak*, Nijmegen: NovioConsult.

Offe, C. (1996), *Modernity and The State*, Cambridge: Polity Press.

Pestman, P. and J. van Tatenhove (1998), Reflexieve beleidsvoering voor milieu, ruimtelijke ordening en infrastructuur: Nieuwe initiatieven nader beschouwd, *Beleidswetenschap*, 12(3), pp. 254-272.

Van der Cammen, H. and L.A. de Klerk (1993), *Ruimtelijke ordening: van plannen komen plannen, de ontwikkelingsgang van de ruimtelijke ordening in Nederland*, Utrecht: Het Spectrum.

Van der Vlist, M. (1998), *Duurzaamheid als planningsopgave. Gebiedsgerichte afstemming tussen de ruimtelijke ordening, het milieubeleid en het waterhuishoudkundig beleid voor het landelijk gebied*, Wageningen: LUW.

Van Schendelen, M. (1997), *Natuur en ruimtelijke ordening*, Amsterdam: NAi.

Van Tatenhove, J. and L. van den Aarsen (1996), Politieke modernisering en doelgroepenbeleid voor het landelijk gebied, *Tijdschrift voor Sociaal Wetenschappelijk ondezoek van de Landbouw (TSL)*, 11 (1996)(4), pp. 253-274.

Van Tatenhove, J. (1999), Political Modernisation and the Institutionalisation of Environmental Policy, Wissenburg, M., G. Orhan and U. Collier eds., *European Discourses on Environmental Policy*, Aldershot: Ashgate, pp. 59-78.

Wetenschappelijke Raad voor het Regeringsbeleid (1998), *Ruimtelijke ontwikkelingspolitiek*, Den Haag: SDU.

Wisserhof, J. and L.C. Crijns (1992), *Ontwikkelingen in integraal waterbeheer; verkenning van beleid, beheer en onderzoek*, Delft: Delft University Press.

Zonneveld, W. (1991), *Conceptvorming in de ruimtelijke planning. Patronen en processen*, Amsterdam: PDI.

8. Agricultural Policy Making in The Netherlands: Beyond Corporatist Policy Arrangements?

JOHAN WISSERHOF

8.1 INTRODUCTION

The domain of agricultural policy offers an interesting field of study for types of policy arrangements and the transformation of one type of arrangement into another. The intensity and scale of production of modern agriculture have contributed greatly to environmental problems such as eutrophication, acidification and the degradation of ecosystems. But in many countries agricultural policies are now being changed and broadened to include concern about environmental pollution by agriculture. In the European Union, the previously productivist Common Agricultural Policy is gradually being reformed into an integrated rural development policy. Due to their radical nature such policy changes are not accomplished easily. They involve conflicts of interest, lengthy negotiations, political struggles, intricate compromises and so forth, which are then institutionalised in new policy arrangements while traditional arrangements may decline.

The aim of this chapter is to investigate which types of policy arrangements have structured the development of agricultural policy over the years, and how transformations of policy arrangements are achieved. The investigation is limited to the Netherlands, a country with a strong agricultural tradition. The development of agricultural policy in the Netherlands is analysed on the basis of the theoretical framework developed in chapter 1.

The primary objective is to see whether an analysis in terms of phases of political modernisation and types of policy arrangements is empirically valid, in the sense that the changes in the agricultural policy domain can be interpreted in a meaningful way. Secondly, the chapter should give an insight into the mechanics of the transformation of policy arrangements, in order to contribute to a further elaboration of the theoretical framework.

The chapter is written primarily from a diachronic perspective. The focus is on the development of agricultural policy in the Netherlands since the Second World War. First the general development of the policy field is described in terms of the political modernisation of agricultural policy, and this is followed by an analysis of the types of policy arrangements that have been formed and transformed.

At the end of the analysis a synchronic perspective is adopted as well, by briefly considering the current coexistence of different types of agricultural and rural policy arrangements. The chapter concludes with the main findings and a discussion of the policy arrangements in the context of the Dutch political system.

8.2 POLITICAL MODERNISATION OF AGRICULTURAL POLICY

The development of agricultural policy in the Netherlands after the Second World War may be divided into three periods: 1945-1980, the 1980s and the 1990s (Bekke et al., 1994).

The first period is characterised by the modernisation of the agricultural sector and by organisational stability in the policy institutions. The 1980s show a growing concern about the environmental impacts of modern agriculture, accompanied by institutional changes. The institutional structures built up in the previous decades were put under pressure from various sides. This second period may be characterised as a turning point in the agricultural policy domain. In the 1990s, the changes set in motion in the 1980s culminated in a reorientation and reorganisation of the field. The previous 'growth model' gave way to a 'sustainable development model', and the Ministry of Agriculture was transformed from a sectoral ministry into a ministry of general administration. These changes can be interpreted as follows in terms of the phases of political modernisation expounded in chapter 1:

- 1945-1980: early political modernisation;
- 1980s: anti-modern criticism; and
- 1990s: late political modernisation.

In other words, the macro-process of political modernisation is reflected quite accurately in the development of agricultural policy in the Netherlands, as will be shown in detail below. To put it the other way round, the process of political modernisation is 'grounded' in the development of Dutch agricultural policy.

8.2.1 Early political modernisation: 1945-1980

After the Second World War food supply was an important concern in the Netherlands, as in many other countries. Growth in agricultural production was a major political priority as part of the general restoration of society. Related priorities were adequate income for farmers and an efficient farming structure (Bekke et al., 1994: 33). Farmers' incomes lagged behind those in industry and services due to the rapid expansion of the latter sectors in the Netherlands after the war. The fragmented farming structure hampered growth in agricultural production as well. The expansion of industry and services also contributed to the expansion of primary production, however. Large numbers of agricultural workers who were no longer needed due to the mechanisation of agriculture could be employed in the other sectors (Bieleman, 1992: 356).

A mix of policy strategies was used to attain the related political priorities of production growth, farmers' incomes and the farming structure. A crucial tactic was the government's financial support for agricultural research, communication and education, the so-called 'OVO-triad' (*Onderzoek* - research, *Voorlichting* - communication, *Onderwijs* - education). Scientific research into agricultural production was strongly promoted, the results were communicated to the farmers through an extensive information service, and the farmers, in turn, were encouraged to attend higher education so as to be able to apply the scientific insights in practice. The agricultural sector in the Netherlands has its own system of education, consisting of primary and secondary agricultural schools and agricultural colleges, culminating in the Wageningen Agricultural University, all under the auspices of the Ministry of Agriculture (Hoetjes, 1993: 109). The government also supported close co-operation between primary agriculture and the agribusiness, which gave strong stimuli to the technical improvement, standardisation and expansion of agricultural production (Bieleman, 1992: 355). The farmers' incomes were guaranteed by means of a detailed market and price policy, which after 1958 at international level was continued in the European Economic Community's Common Agricultural Policy (Bekke et al., 1994: 33). The main policy strategy adopted to rationalise the farming structure was re-allotment and drainage of agricultural land. A large number of land consolidation projects resulted in a radical physical transformation of rural areas in the Netherlands, in order to extensively reorganise and modernise Dutch agriculture (Hoetjes, 1993: 111).

The postwar agricultural policy outlined above fits into the context of early political modernisation. The state was the main policy maker, as it had been since the economic crisis of the 1930s (Hoetjes, 1993: 109), and was supposed to be sufficiently powerful to develop agriculture as a sector of Dutch society. The productivist agricultural policy, moreover, was considered politically non-controversial. The aim was the technical and economic development of Dutch agriculture, nationally and internationally. All political parties supported the policy (Bekke et al., 1994: 34), so ideology and political struggles were not a concern, and possible economic conflicts between different agricultural sectors or rural regions were settled through negotiation and compromise (Hoetjes, 1993: 111). Policy makers and politicians adhered to a technological optimism; technological development was a major means of implementing the policy.

8.2.2 Anti-modern criticism: the 1980s

In the course of the 1970s, the postwar agricultural policy attracted growing opposition from the public. Overproduction, environmental pollution, loss of nature and valuable landscapes, high government expenditure, poor quality of agricultural products, unethical treatment of animals and ecological risks of biotechnology all appeared to be 'side effects' of a one-dimensional productivist policy. In 1980, the new agriculture minister indicated to the parliament that the limits of this policy had been reached, and that a stronger market orientation of the agricultural sector would be necessary. He also focused attention on the importance of nature policy as well as agricultural policy in rural areas (Bekke et al., 1994: 39).

The problems mentioned above required a change in agricultural policy. The motto of the 1950s to 'produce, specialise and rationalise' had to be replaced by control and even reduction of agricultural production (Bieleman, 1992: 357). This was not only a task for the Dutch national government. The problem of overproduction and accompanying budgetary problems received political attention at European Community level as well (Bekke et al., 1994: 39). This eventually resulted in a superduty on milk in order to reduce the surpluses of milk and butter in the EC. At national level, policy efforts concentrated on the problem of excess manure, which is a major concern in contemporary Dutch agriculture (Bieleman, 1992: 356; Frouws, 1993). In 1984, an interim bill was passed through parliament which limited the expansion of pig and poultry farming in the Netherlands. Slightly earlier, in 1982, the Department of Nature Conservation and Outdoor Recreation had been transferred from the Ministry of Culture, Recreation and Social Work to the Ministry of Agriculture and Fisheries. This stimulated the search for a way to integrate agriculture and nature (Bekke et al., 1994: 38). In 1985 the land consolidation act was amended to become a multi-functional land development act. Legally, agriculture was no longer the only type of land use to be considered; land development projects should also include other uses such as nature and recreation. The 1980s, therefore, yielded significant pieces of restrictive and broadened agricultural policy. There were no fundamental policy changes, however, because agricultural policy kept its productivist orientation. Due to technological innovation, production capacity stabilised or even increased.

The changes in agricultural policy in the 1980s cannot be interpreted as political anti-modernisation. The policy changes were not induced from within the agricultural policy domain itself, but mainly by public protests against the effects and features of modern agriculture. Since the 1970s, modern agricultural had come under attack from the environmental movement (Frouws and Van Tatenhove, 1993: 228). As a consequence policy makers were confronted with increasing environmental concern among top politicians and in the parliament (cf. Bekke et al., 1994: 39). But significant changes in the technological emphasis of agricultural policy and its productivist perspective did not occur until the 1990s.

8.2.3 Late political modernisation: the 1990s

The changes set in motion in the 1980s were consolidated and further developed in the 1990s. Essentially, the policy issues of the 1990s do not differ from those of the 1980s, but they became clearer and were discussed more openly (Bekke et al., 1994: 66). The completion of the GATT Uruguay Round in 1993, for example, established the ongoing liberalisation of agriculture at global level. The workings of the market rather than protectionist public policies increasingly drive agricultural production in its various forms. The importance of consumer demands is also increasingly being recognised in agricultural policy. These demands, in industrialised countries, have gradually begun to point in the direction of high-quality food rather than bulk production. Related issues are the continuing ethical objections to factory farming, and biotechnology. The issue of the relationship between agriculture and nature, finally, has become firmly established on the political agenda. Both public opinion and political pressure strongly advocate the integration of agriculture and nature (Bekke et al, 1994: 67), possibly in combination with recreational and residential use.

Differences between the 1980s and the 1990s can be found in agricultural policy itself and in the widening of agricultural policy into rural development policy. From the early 1990s onwards agricultural policy was severely tightened (Frouws and Van Tatenhove, 1993: 231). A number of government decrees dealt with manure surpluses, acidification, pollution from pesticides and heavy metals, drying out of the soil, farm wastes, etc. The agri-environmental policy objectives were reformulated in plans such as the Structure Plan for Agriculture and the Multi-Year Crop Protection Plan, to include considerable reductions in environmental pollution through farming practices. Attention shifted from production to pollution (Frouws and Van Tatenhove, 1993: 228).

The most significant change compared to the 1980s, however, is that agricultural policy was broadened to become an integrated rural development policy. The policy field of the Ministry of Agriculture was widened, dealing not only with primary production but with rural land use in general (cf. Van Tatenhove, 1993: 300). This change in scope was consolidated in the 1992 national Structure Plan for the Rural Areas in the Netherlands. In this plan, several region-specific projects such as Rural Area Pilot Projects and Valuable Man-Made Landscapes were presented, to complement the Area-based Integrated Spatial Planning and Environment Projects launched by the Ministry of the Environment (cf. Frouws and Van Tatenhove, 1993). These projects comprise a variety of aspects of rural life: agriculture, environment, nature, recreation and so on. Specific agricultural policy, aimed at a stronger market orientation and the restructuring of agriculture, was set within the wider context of a 'renewal' of the countryside. Farmers were encouraged to apply new environmental technologies, to develop regional quality products, to do business in tourism or nature management, and so forth in order to sustain the countryside. The role of the Ministry of Agriculture and Fisheries (which was renamed the Ministry of Agriculture, *Nature Management* and Fisheries in 1989) was changing from that of a sectoral client ministry to becoming a ministry of general administration (Frouws and Van Tatenhove, 1993: 224; Bekke et al., 1994).

The widening of agricultural policy into rural development policy is indicative of late political modernisation. The region-specific projects include strong public participation in policy making, not only by farmers' organisations but also by environmental groups, recreational organisations, rural inhabitants and others. Moreover, the initiative for these projects often comes from regional government rather than the state, although the region-specific approach as such was developed at national level. The state is just one of the actors operating in the regional network which functions as a governing body. The region-specific approach thus involves a decentralisation of policy making. 'Bottom-up' processes of self-regulation at regional level (Frouws and Van Tatenhove, 1993: 237) replace the state-dominated central rule approach to policy making in early political modernisation.

8.3 TRANSFORMATION OF AGRICULTURAL POLICY ARRANGEMENTS

The process of political modernisation of postwar agricultural policy in the Netherlands resulted in the institutionalisation of different types of policy arrangements, ranging from corporatist to a plurality of corporatist, statist, liberal and civic policy arrangements. These policy arrangements are briefly typified in table 8.1 with regard to their governing coalition(s) and central discourse. Below, the rules and resources of the various arrangements are discussed as well.

policy arrangement	*Governing coalition(s)*	*central discourse*
corporatist	'Green Front'	'growth' and 'rationalisation'
Statist	Ministries of Agriculture and the Environment	'limits to growth'
Civic	Regional networks	'sustainable development' and 'liveability'
Liberal	Market players at (inter)national level(s)	'free trade'

Table 8.1: Typification of agricultural policy arrangements

8.3.1 Early political modernisation: corporatist and statist policy arrangements

In the phase of early political modernisation, the dominant policy arrangement is corporatist in nature. A corporatist policy arrangement may be defined as

> 'a socio-political structure of interest articulation and policy formation, in which functional interest organisations (farmers' unions) possess a representational monopoly, co-operating between each other and with the state on the basis of political-economic consensus at the top. The participating organisations are granted privileged influence on public policy-making in exchange for disciplining their constituency (the farmers) and restraining their demands' (Frouws and Van Tatenhove, 1993: 223; cf. Frouws, 1993: 47).

A corporatist policy arrangement, therefore, may be considered an institutionalised exchange or 'package deal' between the state and a specific socio-economic sector, in this case agriculture. The components of this 'deal' are summarised in table 8.2.

Rules:
Consensus, elitism, technocracy, de-politicisation, insulation
Resources:
State: influence, authority, information, monopoly
Interest organisations: co-operation, information, discipline, legitimacy

Table 8.2: The corporatist policy arrangement (cf. Frouws, 1993: 43; 1996: 32)

A precondition for such a 'package deal' is consensus on the main features of agricultural policy, in other words the policy discourse. In the phase of early political modernisation the policy discourse was indeed undisputed. As indicated in table 1 and described earlier in this chapter, growth in agricultural production was a major political priority after the Second World War. This was understood as a technological rather than a political issue, and therefore specialist expertise and policy-making efficiency were considered more important than democratic policy-making procedures. This contributed to the relative insulation of the postwar agricultural policy community, the 'Green Front'. It consisted of three core parties: the state (represented by the Ministry of Agriculture), the farmers' unions (represented by the Board of Agriculture, the *Landbouwschap*, at that time the co-ordinating organisation for the national farmers' unions and the agricultural trade unions), and the parliamentary committee on agriculture. Within this 'Iron Triangle', the objective to develop a modern, competitive, export-oriented agricultural sector was rarely questioned. The 'gentlemen's agreements' reached by top officials of the ministry and executives of the Board were 'smoothly legitimated' by the parliamentary committee. Members of parliament not belonging to this committee, other ministries and non-agricultural interest groups could barely penetrate the 'Iron Triangle' (Frouws and Van Tatenhove, 1993: 222). In short, the elite of the Green Front governed agricultural policy making.

To put the corporatist arrangement of postwar agricultural policy making in a wider social and political context, important reasons for its rise and continuance were the agricultural crisis of the 1930s and the idea of 'agrarian particularism'. In the 1930s the national government strongly intervened in the agricultural sector through a number of laws (on wheat, dairy products, pork and farm rents, culminating in an Agricultural Crisis Act in 1933/34), in order to protect the farmers from the effects of a period of overproduction after the First World War (Bieleman, 1992: 222-227). Thus, the basis was laid for the collaboration between the state and agricultural interest organisations which is typical of the corporatist policy arrangement. After the Second World War, the political priority of food production brought about a renewal of the close co-operation between the state and the agricultural sector. The prewar Ministry of Agriculture and Fisheries was re-established (under the new name of Ministry of Agriculture, Fisheries and Food Supply) (Bieleman, 1992: 237), and its energetic first minister, Sicco Mansholt, institutionalised the co-operation with the agricultural sector by recognising the Association for Agriculture (the later Board of Agriculture) as a representative of the sector and by establishing regular monthly meetings between the ministry and the Association (Louwes, 1980: 226). Due to the changing political relationships after the war, government intervention in the agricultural sector took on a supranational dimension as well, in the EEC's Common Agricultural Policy. Interestingly, Sicco Mansholt became the first EEC Agriculture Commissioner.

Agriculture is the only sector of the Dutch economy in which corporatism has been a long-lasting policy arrangement. Its roots in the prewar agricultural crisis alone cannot explain this. An important additional explanatory factor is the ideology of 'agrarian particularism' (Frouws, 1993: 73-74; 1996: 40-41). In many European countries, the presumed particularity of agriculture has legitimated a strong government involvement in the sector (Frouws, 1993: 71; cf. Hoetjes, 1993: 118). Main 'particularities' are the agricultural structure of small family firms, the relative uncertainty and inflexibility of primary production, the low price elasticity of food and the paramount importance of a regular food supply. The fact that most farms are relatively small businesses means that they cannot provide for the preconditions of the modernisation of agriculture (e.g. research and development). The fact that they are mostly family firms means that large-scale bankruptcies could lead to considerable social unrest and threaten the stability of society as a whole. The relative uncertainty of primary production due to climatic conditions was considered a justification for governmental support as well, while the inflexibility of primary production does not allow farmers to respond quickly to price fluctuations. Combined with the low price elasticity of food, this could lead to a lasting fall in food prices that would spell the end for many farms. The paramount importance of a regular food supply, finally, completes the ideology of agrarian particularism, together with the political wish for a 'native' (national or European) food supply. The presumed particularity of agriculture has led to the idea of a 'corporatist imperative' in this sector (Keeler, 1987).

We shall now investigate what remained of the 'corporatist imperative' after the anti-modern criticism of the 1980s. The 1970s already saw growing public and political concern for the environment, fuelled by an increasing body of ecological knowledge. However, agricultural pollution as yet attracted little public attention. The Arcadian notion of the farmers' intimate relationship with nature still prevailed, accompanied by the idea that the agricultural sector was one of the driving forces of the Dutch economy. Only particular aspects of intensive livestock farming such as its smell and its treatment of animals provoked public indignation (Frouws and Van Tatenhove, 1993: 228). It was only in the 1980s that agri-environmental problems became an issue of serious public and political concern. This was partly due to the near monopoly of the Green Front on knowledge and information about the agricultural sector and farming practices (the 'OVO'-triad). Research into possible adverse effects of agriculture was not encouraged; production was much more important than pollution. In the 1980s, however, the adverse effects of intensive agricultural production gradually began to be put forward by agricultural researchers, as well as by biologists, drinking-water experts and environmental scientists, all of whom rejected the dominant orientation of productivist research.

The growing number of research reports on environmental pollution through agriculture, and on the increase in the number of pigs in the Netherlands, put ever greater political pressure on the Ministry of Agriculture to change its policy (Frouws, 1996: 35). As was indicated earlier, the new agriculture minister at that time, Gerrit Braks, was sensitive to the political signals, and openly admitted that the limits of the productivist policy had been reached. In 1982, he reached agreement with the energetic new environment minister, Pieter Winsemius, on legislating on manure. This agreement was historic in the sense that the Ministry of the Environment, after fifteen years of interdepartmental struggles, was allowed a say in the preparation of this legislation: an 'outsider' entered the Green Front (cf. Frouws, 1993: 82-85). Due to the political priority of the 'environment', the environmental ministry was expanding considerably. From 1972 to 1992 the number of employees increased from 150 for the ministry as a whole to 1,160 for the Department of the Environment alone (Bekke et al., 1994: 49). In the 1980s the ministry could therefore afford a more offensive strategy than before. The result of this offensive was the previously mentioned 1984 interim bill on pig and poultry farming, which limited the further growth of intensive livestock farming. The contents of the bill had not been negotiated with the farmers' unions. It was prepared in secret by a small group of top officials and lawyers from the Ministries of Agriculture and the Environment (Bekke et al., 1994: 42). Not surprisingly, the agricultural sector perceived the interim bill (which became law as soon as it was submitted to parliament, in order to prevent an 'announcement effect') (Frouws, 1996: 35; Bekke et al., 1994: 39) as an attack by its 'own' ministry (Frouws, 1993: 92; 1996: 35). One year earlier, in 1983, the postwar productivist agricultural policy had already been modified when the superduty on milk was enforced. This was the implementation of a European directive, however, whereas the interim manure bill was a national initiative. Nevertheless, the superduty meant that the Green Front had encountered yet another 'outsider': the European Community. There had been a long-standing relationship between the Ministry of Agriculture and Fisheries and the EC, of course, but due to the EC's restrictive stance on both agricultural (over) production and fisheries this relationship was now put under pressure (Bekke et al., 1994: 40).

The overall picture of the Ministry of Agriculture and Fisheries in the 1980s shows the already mentioned shift from a client ministry, solely orientated towards the agricultural interest, to a ministry of 'general' administration, protecting the public interest (Hoetjes, 1993: 123). The environmental and budgetary problems of excessive agricultural production put increasing public and political pressure on the ministry. The solid reputation of agriculture as a major socio-economic sector declined. A number of parties outside the Green Front challenged the productivist agricultural policy: the EC, the parliament, the Ministry of the Environment, and nature and environmental organisations. These previously 'peripheral' actors gradually moved to the centre of agricultural policy making (Bekke et al., 1994: 40). Quite significant in this respect was the merging of the Department of Nature Conservation and Outdoor Recreation with the Ministry of Agriculture and Fisheries during the formation of the new government in 1982. A land consolidation project on the island of Terschelling had demonstrated that agriculture and nature were becoming ever more interwoven. Another reason for the reorganisation was that the Ministry of Agriculture was eager to strengthen its political position now that the number of farmers was gradually decreasing (Bekke et al., 1994: 50). As was indicated before, the reorganisation proved to be the catalyst for a widened scope for the ministry. It had to change its exclusive orientation on agricultural production towards a policy that balanced agriculture with nature, landscape, outdoor recreation, environment, food quality and animal welfare (Hoetjes, 1993: 123; Frouws, 1996: 34). In particular the top officials of the ministry sensed the changing ideas at the political level and tried to translate these changes into restrictive policy: the national implementation of the superduty and the interim bill on pig and poultry farming (Bekke et al., 1994: 39).

The changing position of the Ministry of Agriculture alienated the farmers' unions. Suddenly, the ministry appeared not to represent the farmers' interests anymore. Moreover, the unions were not even consulted in policy preparation. This course of events was, understandably, accompanied by misunderstandings, disappointments, conflicts and struggles between the unions and the ministry (Hoetjes, 1993: 124). The unions, in turn, lost their credibility among the farmers themselves. The 'bad news' of the various restrictive laws and regulations made them turn away from their former representatives (Frouws, 1996: 34). Also, their better education, specialisation and technical expertise brought about a more critical attitude towards the unions than had existed before.

To sum up, the tight links in the 'Iron Triangle' (Ministry of Agriculture - Board of Agriculture - parliamentary committee on agriculture) were successively broken. First that between the ministry and the parliamentary committee, then that between the ministry and the Board, and finally also the representational bond between the Board and the farmers. This gradual decline of the Green Front in the course of the 1980s is roughly schematised in figure 1.

overproduction, environmental pollution
↓ (scientists, journalists)
'green politics' ← nature conservation and outdoor recreation
↓ (top officials)
restrictive agri-environmental policy and legislation
↓
alienation between Ministry of Agriculture and farmers

Figure 8.1: Erosion of the corporatist policy arrangement

Clearly, the centralist corporatist policy arrangement did not withstand anti-modern criticism. Major conditions of the corporatist deal (see table 8.2) were no longer met. There was no consensus between the ministry and the farmers' unions on the policy discourse. Instead of promoting unlimited growth in agricultural production, the ministry was put under public and political pressure to recognise the 'limits to growth'. The new discourse retained the technocratic character of the previous growth discourse, however, but technology was re-orientated, from simply growth in agricultural production to prevention of environmental pollution while maintaining or even extending production (Frouws and Van Tatenhove, 1993: 229). On the other hand, agricultural policy was no longer a depoliticised technological issue, but became a subject of political debate and interdepartmental struggle. Faced with of the EC, the parliament, the Ministry of the Environment, and nature and environment groups, the insulation of the corporatist policy arrangement could not be maintained. Its elitism was also outdated because of the break between the farmers and their previous representatives.

Curiously, in response to anti-modern criticism agri-environmental policy making adopted a statist arrangement, whereas one might have expected a democratisation of policy making. The 'political demodernisation' following the 'erosion of the corporatist partnership' was accompanied by 'statist regulation' (Frouws and Mol, 1997: 284). The main rules and resources of the statist policy arrangement are summarised in table 8.3, on the basis of the above analysis.

Rules:
'general' administration, autocracy, secrecy
Resources:
ecological knowledge, legal expertise, legislative power

Table 8.3: The statist policy arrangement

Policies were prepared in secret by a small group of top officials and lawyers in the Ministries of Agriculture and the Environment, as was described above. In other words, the state governed policy making. The policy changes as such were, of course, a result of public and political pressure on the state, but the earlier participation by farmers in policy making was not replaced by a wider participation by civil society and the market. The corporatist policy arrangement had involved a close relationship between the state (Ministry of Agriculture), civil society (parliamentary committee on agriculture) and the market (Board of Agriculture). Now, the state became the dominant policy maker.

Despite the overall decline of the corporatist agricultural policy arrangement in the 1980s, the Green Front still maintained a significant position of power. This is seen first of all in the fact that the adverse environmental effects of intensive agriculture became a political issue only during the course of the 1980s, while these effects had already been reported in the 1970s. Furthermore, the 1984 interim bill on pig and poultry farming did not directly limit the number of animals - the number has actually increased significantly - but only indirectly, through regulating the quantity of manure that could be produced. The Green Front was able to establish high manure absorption standards for agricultural soil, initially to the extent that there would be no manure surplus at all at national level. Moreover, the sector was given fifteen years to solve the manure problem (Frouws, 1996: 36). In other words, the agricultural interest still strongly influenced policy making. The Ministry of Agriculture remained the champion of the agricultural interest in the preparation of legislation on manure (Frouws, 1993: 95). The tight institutional link at top level between the state, civil society and the market in the agricultural sector were not abandoned instantly. The later stages of the early political modernisation of agricultural policy, therefore, exhibit a corporatist as well as a statist policy arrangement. Only in the phase of late political modernisation was the central corporatist arrangement truly eroded, as will be described in the next section.

8.3.2 Late political modernisation: corporatist, statist, civic and liberal arrangements

Despite the policy changes in the 1980s, there had been very few significant changes in the administrative organisation and management style of the Ministry of Agriculture (Bekke et al., 1994: 90-95), which remained strictly hierarchical. Calls for reorganisation were even suppressed. The policy changes were prepared by a small group of top officials, who barely communicated with their subordinates and simply did not consider possible problems of policy implementation by the administrative organisation (Bekke et al., 1994: 55). This traditional approach partly explains why it was possible to make radical policy changes. In an open, communicative organisation with a democratic management style such changes would have been much more difficult to initiate (Bekke et al., 1994: 91).

Nevertheless, anti-modern criticism paved the way for a fundamental reorganisation of the ministry. In the early 1990s, it transformed itself from a rigid and bureaucratic into an enterprising and learning organisation (Bekke et al., 1994: 73). The ministry was almost forced to do so by a number of political events: first the incorporation of the Department of Nature Conservation and Outdoor Recreation; secondly, strong criticism from the media on the inspection of fish quotas, and two negative research reports on the image of the ministry among its target groups; thirdly, an interpellation in the parliament about supposed intimidatory practices that had been adopted within the ministry and the resulting audit of the ministry by an independent commission in 1991. All these incidents eventually led to the start of large-scale reorganisation: the so-called Swallow process (Bekke et al., 1994: 49-55). Core elements of the Swallow process were open internal communication, 'outward' orientation, and delegation of responsibilities (Bekke et al., 1994: 56). These elements fitted in well with the general trend towards decentralisation and the growing importance of the subnational administrative levels since the 1980s (Hoetjes, 1993: 125). In 1990 a 'large efficiency operation' was begun by the national government in order to accomplish a 'retreat of central government' (restraint on the main policy lines, separation of policy making and policy implementation, policy implementation by regional departments, and decentralisation wherever possible (Bekke et al., 1994: 128-129). The Swallow process was perfectly adapted to this operation (Bekke et al., 1994: 73).

The internal reorganisation of the Ministry of Agriculture was reflected in its external position as well. In 1994, it presented a new management approach to agri-environmental problems: 'tailor-made management' (*Sturing op maat*, 1994). In this policy document the ministry refrained from the traditional regulatory approach and opted for a greater role for individual farmers, agribusiness, consumers and environmental organisations. The state would only provide the means for adequate policy implementation and self-regulation by these parties: it would play an 'enabling' role. The policy document was also innovative in the sense that it was not prepared by the Ministry of Agriculture alone, but by a working group which also included representatives of the Ministry of the Environment, agribusiness, environmental organisations and scientists (Bekke et al., 1994: 70-71). The preparation of the document thus demonstrates that the insulated Green Front had been transformed into a 'combined agro-environmental policy network' (Frouws and Van Tatenhove, 1993: 233).

The regional level was crucial in the ministry's new management approach, not only because of the emphasis on policy making by regional departments and decentralisation of governmental responsibilities but also because of the 'regionalisation strategy' pursued by the Ministry of the Environment (Van Tatenhove, 1993: 300). In the late 1980s the environmental ministry launched an 'integrated region-specific approach' in order to improve the poor implementation of environmental policy, not in the least agri-environmental policy. Until then, this ministry had barely been able to influence the environmental impacts of agricultural production due to the strong social and political position of the Green Front. Through the regionalisation strategy, the ministry intended to co-operate with agricultural target groups in so-called Area-based Integrated Spatial and Environment Projects. This strategy combined well with the 'broadening strategy' of the Ministry of Agriculture (Van Tatenhove, 1993), which, moreover, had a long-standing planning tradition at regional level in land development projects. Since the Land Development Act of 1985 these projects could be used to integrate agriculture, recreation, environment, nature and landscape. In the early 1990s the ministry initiated several integrated region-specific projects, such as Rural Area Pilot Projects, Valuable Man-Made Landscapes and National Parks.

Late-modern policy making at regional level involves a civic policy arrangement with new roles for the state, the market and civil society. The state has abandoned its pretension of omnipotence and central steering abilities. It only sets the key objectives and the boundary limits, provides administrative and financial facilities and supervises rural development 'at a distance': the 'enabling' role mentioned earlier. Although it does maintain its normative powers, the state is not the central policy maker any longer. Private actors have become important participants in policy making at regional level (Van der Vlist, 1998). This implies the models of the 'negotiating' and the 'calculating' government, in which the relationships between state, market and civil society are largely informal and participating actors have a more or less autonomous position; public and private parties are highly interdependent (Van Tatenhove, 1993: 301; 94-95). Policy making has become a matter of public-private partnerships. Both public and private parties participate in the regional policy-making networks; the traditional dividing lines between the state, the market and civil society have been crossed.

As a result, integrated region-specific projects involve new roles for the market and civil society as well. Various private actors have a say in regional policy making: individual farmers, agricultural co-operatives, landowners, shopkeepers, etc. The state encourages an active role for these actors in, for example, environmental protection, nature conservation, recreation and tourism, through special 'rural renewal' funds. Although the state itself could play a more active stimulating role in this respect, many farmers argue for self-rule over environment, nature and landscape by the taking over governmental tasks and responsibilities (Horlings, 1996). To that end, they have founded a number of 'environmental co-operatives' or 'innovation groups': associations of farmers aiming for sustainable agriculture, for instance by adapting their cultivation techniques to local ecological conditions. Landowners also claim they are able to implement policy objectives better than government institutions or nature organisations. Especially since the arrival of a politically-liberal Minister of Agriculture, Nature Management and Fisheries in 1994, the ministry is quite responsive to such market initiatives.

Effective public-private partnerships require a common discourse on rural development; the goals of development have to be agreed upon. In fact, region-specific policy making has been concerned more with consensus building on policy objectives and instruments than with regional specifications for national policies (Van Geleuken, 1998). Since the appearance of the notion of sustainable development in the late 1980s, harmonisation of economy and ecology has become a major objective in various policy fields in the Netherlands. Although different organisations adhere to different definitions of the concept, it serves as a guiding principle for policy making in rural areas (cf. Van der Vlist, 1998). Both public and private parties are generally convinced of the need for sustainable agriculture in relation to cultural-historical landscapes, rural communities, biological diversity and aquatic ecosystems. Thus, agricultural production is linked to general issues of rural development. The civic policy arrangements, therefore, embody a clear shift from the corporatist and statist arrangements, not only with regard to their governing coalitions but also in their central discourse. The major rules and resources of civic policy arrangements are summarised in table 8.4.

Rules:
Consensus, depoliticisation, egalitarianism, public-private partnership
Resources:
Knowledge and information, managerial and communicative skills, finances

Table 8.4: Civic policy arrangements

Although civic policy arrangements appear to be dominant in the phase of late political modernisation, they are not exclusive. Even the civic arrangements themselves may have corporatist elements if the participants do not truly represent their constituency. In particular farmers' organisations often lack the full support of all individual farmers in a region nowadays, due to the ongoing differences in styles of farming, each style representing a specific structure of the labour process and of farm development. As a result the agricultural interest organisations can no longer speak for all farmers. This means that their representational monopoly - an important resource in the corporatist policy arrangement - does not exist any longer, and it also qualifies the presumed egalitarianism of the civic arrangements. If the actors participating in a regional policy network do not actually represent all stakeholders, the network resembles an insulated corporatist elite. Other characteristics of the corporatist policy arrangement also reoccur in the civic arrangements, such as a desire for consensus and the depoliticisation of policy making, as will be discussed in the conclusion to this chapter.

At (inter)national level, liberal (market-oriented) policy arrangements are formed because of pressure from the 1993 GATT treaty and the 1992 MacSharry reforms of the EU's Common Agricultural Policy. The Ministry of Agriculture strongly encourages market-oriented agricultural production. Consumer demand for food should determine agricultural production. The sector itself is given the responsibility of adapting its production to consumer demand (self-regulation). Government price support for specific agricultural products has been stopped or decreased. Major resources in maintaining the competitive advantages of Dutch agriculture are (food and environmental) knowledge and technology. These should be applied in innovate types of agricultural production. The introduction of quality standards is also an important resource for the strong market position of Dutch agriculture (LNV, 1995: 11-12). The espousal of competitive market solutions for agricultural production, and by implication the dismantling of much of the state regulation of agriculture, seems to be a 'nearly universal trend' (Vail, 1994: 73). The main rules and resources of liberal policy arrangements are summarised in table 8.5.

Rules:
Market orientation, less government intervention, self-regulation
Resources:
Knowledge and technology (innovation), quality standards (certification)

Table 8.5: Liberal policy arrangements

At national level various types of policy arrangements exist simultaneously. Since the decline of the Green Front, the articulation of agricultural interests has been sectoralised and regionalised. Separate agricultural sectors have established their own interest organisation or trade union. Some of these, such as the horticultural interest organisation, have close contacts with the Ministry of Agriculture and the ongoing restructuring of horticulture was developed in close co-operation with the sector itself. The horticultural farmers are relatively strongly market-oriented, which fits in well with current agricultural policy. Also, they are concerned about their environmental image among the public and are quite ready to invest in environmental technology. In short, the relationship between the state and the horticultural sector resembles a corporatist policy arrangement. The pig farming sector, on the other hand, has a much more militant relationship with the state. Most of the sector strongly disagrees with the state's agri-environmental policy. Unlike the horticultural sector, interest articulation in the pig farming sector occurs via a variety of farmers' organisations. It is probably due to the antagonist relationship that policy making for the pig farming sector has a statist arrangement. Both in policy on manure and in the recent restructuring of the pig farming sector has the state followed a central rule rather than a network management or a self-regulation approach. Another example of a statist policy arrangement can be found in nature policy, namely the preparation of the 'Primary Ecological Infrastructure'. The plan for this ecological infrastructure was primarily prepared by the state. The implementation of the National Ecological Network, however, is an example of a civic arrangement, because the Ministry of Agriculture is dependent on the willingness of landowners and farmers to co-operate. All it can do is mobilise financial resources to influence decisions of these actors (output steering on defined nature types) (De Jong and Van Tatenhove, 1998: 167, 172). Similarly, the implementation of the reconstruction of pig farming areas requires a modification of the aforementioned statist arrangement. This arrangement is now complemented by regional initiatives, in civic arrangements of integrated region-specific projects.

To sum up, the phase of late political modernisation does not exhibit a uniform arrangement of policy making, but shows the coexistence of a plurality of policy arrangements.

8.4 CONCLUSIONS AND DISCUSSION

It may be concluded that the theoretical framework is empirically valid. The respective phases of political modernisation (early, anti- and late) can be empirically identified in the development of agricultural policy in the Netherlands, as can different types of policy arrangements (corporatist, statist, civic and liberal). There is no straightforward relationship, however, between the phases of political modernisation and the types of policy arrangements. Different policy arrangements may coexist in any phase of political modernisation. The analysis above pointed out the existence of both corporatist and statist arrangements in the phase of early political modernisation and the coexistence of all four types of policy arrangements at different administrative levels in the phase of late political modernisation. Also, a rather curious development was found, namely the formation of a statist policy arrangement in response to anti-modern criticism of agricultural policy.

Macro-processes of social and political change drive the transformation of one type of policy arrangement into another on the one hand, and day-to-day policy making on the other. The two world wars were significant macro-processes, and they explain the productivist orientation of 'postwar' agricultural policy as well as the strong government intervention on agricultural production and the resulting emergence of a corporatist policy arrangement. The ideology of 'agrarian particularism' also appeared to be an important macro-condition in this respect. Important macro-processes explaining the decline of the corporatist policy arrangement are: the rise of anti-modern criticism with regard to environmental pollution in the 1970s, as part of the increasing consideration of non-material values in industrialised societies; and the institutionalisation of these concerns in the expanding number of public interest organisations as well as in the environment ministry in the 1980s. Another significant process at the macro-level in this regard is the global shift from Fordist mass production to a production regime of specialisation, differentiation and sectoralisation (Bekke et al., 1994: 18-19). The EC's restrictive attitude towards both agricultural (over) production and fisheries may also be considered an explanatory factor. The rise of 'new right' politics in the 1980s, finally, resulting in the 'large efficiency operation' by the national government in the Netherlands, is a macro-process which paved the way for the emergence of civic policy arrangements in the phase of late political modernisation.

Within the context of these structural developments several agents played important roles in the transformation of agricultural policy arrangements in day-to-day policy making. Scientists, for example, highlighted the characteristics and scale of environmental pollution through agriculture. Journalists focused public attention on the 'ecological crisis' in the 1970s and on the management problems of the Ministry of Agriculture in the 1980s. Members of parliament who made an interpellation about the supposed intimidatory practices within the ministry forced the management problems into the political limelight. Ministers were found to have played important roles as well. A new, energetic minister was an influential agent in both the emergence of the corporatist policy arrangement (Sicco Mansholt: agriculture) and the decline of this arrangement (Pieter Winsemius: environment). The awareness of the new agriculture minister in the early 1980s (Gerrit Braks) of the limitations to the productivist agricultural policy is another factor explaining the decline of the corporatist policy arrangement. Top officials at the Ministry of Agriculture also played an important part, in translating the public and political concerns into restrictive agri-environmental policy and legislation. The arrival of a liberal agriculture minister in the 1990s, finally, contributed to the emphasis on market orientation in current agricultural policy and to the stronger role of market parties in rural areas.

The theoretical framework has highlighted two developments with regard to agricultural policy arrangements in the Netherlands:

- broadening, from the elite of the 'Green Front' to multiple-issue networks; and
- decentralisation, from national to regional level (figure 8.2).

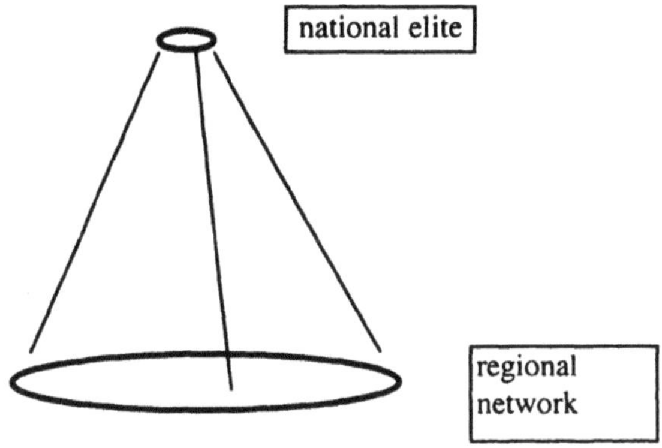

Figure 8.2: Broadening and regionalisation of agricultural policy making

The elite policy making at national level in the postwar corporatist policy arrangement has been replaced by civic policy arrangements in which both public and private parties participate. Policy making has become a matter of public-private partnerships. The civic arrangements embody a change of rules and resources of policy making, compared to the corporatist arrangement (tables 8.2 and 8.4). Important rules which have changed are the switch from elitism to egalitarianism, from insulation to networking, and from a somewhat autocratic to a more democratic method of policy making. Main resources in the civic policy arrangements are managerial and communicative skills and finances, while the corporatist resources of authority, monopoly and discipline have lost their significance.

Despite these changes, civic policy arrangements show important similarities with the corporatist arrangement. Prominent rules in both types of policy arrangements are the focus on consensus and the depoliticisation of policy making; knowledge and information remain important resources of policy making. These characteristics may be considered typical of the Dutch political system (cf. Andeweg and Irwin, 1993: 229-240). Although the Dutch have been considerably 'consociated' (*verzuild*) in the past, each consociation having its own ideology, there has always been a need for consensus regarding functional activities, not in the least because of the common threat of the sea. Consequently, politics in the Netherlands is rather pragmatic. Policy making is characterised by consensus building and the avoidance of ideological controversies. A heavy reliance on factual information as a major resource of policy making fits well into these rules. As only specialists are able to use this information adequately, they have a strong influence on policy making in the Netherlands. Specialists in a given policy field have an important position within the ministries. Both the corporatist and the civic policy arrangements are staffed and managed by experts in the field of agriculture and rural development. In these respects, the current civic policy arrangements do not differ from the corporatist arrangement. Representatives of state, market and civil society are all represented in both types of policy arrangements. The number of participants has increased, but the nature and the aim of participation remain the same: consensus on the basis of factual information.

In the light of this analysis, the statist agri-environmental policy arrangement during the 1980s may be considered an 'un-Dutch' interlude. It would not have been necessary if the Green Front, including the Ministry of Agriculture, had been more sensitive to public and political opinion. The statist arrangement served to swiftly adapt agricultural policy to the changing socio-political circumstances, but only to continue previous (corporatist) consensus-building policy making on a wider range of issues and at a regional level in civic arrangements. Anti-modern criticism paved the way for the institutionalisation of a mixture of policy arrangements in the phase of late political modernisation.

Combined with the fact that the corporatist arrangement has persisted at national level in some sections of the agricultural sector until the present day, we may conclude that agricultural policy making in the Netherlands has not in fact progressed beyond corporatism. In the light of the international reputation of the corporatist Dutch 'polder model', this might well be the strength rather than a weakness of this policy domain.

REFERENCES

Andeweg, R.B. and G.A. Irwin (1993), *Dutch Government and Politics*, Basingstoke: Macmillan.

Bekke, H., and J. de Vries (1994), *De salto mortale van het Ministerie van Landbouw, Natuurbeheer en Visserij*, Alphen a/d Rijn: Samsom H.D. Tjeenk Willink.

Bieleman, J. (1992), *Geschiedenis van de landbouw in Nederland 1500-1950: Veranderingen en verscheidenheid*, Meppel/Amsterdam: Boom.

De Jong, D. and J. Van Tatenhove (1998), The Institutionalization of Dutch-German Cross-Boundary Nature Policy, *Sociologia Ruralis*, 38(2), pp. 163-177.

Frouws, J. (1993), *Mest en macht. Een politiek-sociologische studie naar belangenbehartiging en beleidsvorming inzake de mestproblematiek in Nederland vanaf 1970*, Wageningen.

Frouws, J. (1996), Politieke modernisering van het groene Front, *de Sociologische Gids*, XLIII(1), pp. 30-45.

Frouws, J. and A.P.J. Mol (1997), Ecological Modernization Theory and Agricultural Reform, H. d. Haan and N. Long, eds., *Images and Realities of Rural Life*, Assen: Van Gorcum, pp. 269-286.

Frouws, J. and J. v. Tatenhove (1993), Agriculture, Environment and the State. The Development of Agro-environmental Policy-making in the Netherlands, *Sociologia Ruralis*, XXXIII(2), pp. 220-239.

Hoetjes, B.J.S. (1993), *Landbouw, Natuurbeheer en Visserij: profiel van een ministerie*, 's-Gravenhage: VUGA.

Horlings, I. (1996), *Duurzaam boeren met beleid. Innovatiegroepen in de Nederlandse landbouw*, Nijmegen, Wageningen.

Keeler, J.T.S. (1987), *The Politics of Corporatism in France: Farmers, the State and Agricultural Policy-Making in the Fifth Republic*, Oxford: Oxford University Press.

Louwes, S.L. (1980), Het gouden tijdperk van het groene front; het landbouwbeleid in de na-oorlogse periode, G.A. Kooy, J.H. de Ru and H.J. Scheffer, eds., *Nederland na 1945*, Deventer: Van Loghum Slaterus, pp. 223-249.

Ministerie van Landbouw, Natuurbeheer en Visserij (1995), *Dynamiek en vernieuwing*, Den Haag: SDU.

Vail, D. (1994), Sweden's 1990 Food Policy Reform: From Democratic Corporatism to Neo-liberalism, P. McMichael, eds., *The Global Restructuring of Agro-Food Systems*, Ithaca/London: Cornell University Press, pp. 53-75.

Van der Vlist, M. (1998), *Duurzaamheid als planningsopgave. Gebiedsgerichte afstemming tussen de ruimtelijke ordening, het milieubeleid en het waterhuishoudkundig beleid voor het landelijk gebied*, Wageningen: LUW.

Van Geleuken, B.P. (1998), *Gebiedsgericht grondwaterbeschermingsbeleid. De betekenis van congruentie*, Nijmegen.

Van Tatenhove, J. (1993), *Milieubeleid onder dak? Beleidsvoeringsprocessen in het Nederlandse milieubeleid in de periode 1979-1990; nader uitgewerkt voor de Gelderse Vallei*, Wageningen: Pudoc.

9 Conclusions and Research Agenda: Political Modernisation and the Dynamics of Environmental Policy Arrangements

BAS ARTS, JAN VAN TATENHOVE AND PIETER LEROY

9.1 INTRODUCTION

Apart from its aim to formulate the general conclusions of the book, this final chapter is at the same time intended as a stimulus for further discussion and research. We hope to achieve this by combining and comparing the two parts of this book. Part I dealt with theories on institutionalisation, political modernisation and policy arrangements. Based upon these theories and insights, we developed a specific conceptual framework to analyse environmental policy making, centred on the 'policy arrangements' concept. A 'policy arrangement' refers to the temporary stabilisation of the organisation and substance of a policy domain at a specific level of policy making. By introducing this concept we believe it is possible to analyse the dynamics of environmental politics as being the interplay or duality between macro-processes of social and political change, understood as political modernisation, and interactions between actors in day-to-day policy processes. In part II of the book five authors applied this analytical scheme to specific domains or cases in environmental policies.

In comparing and combining these two parts, we hope to achieve the double goal mentioned above. First, we will try to answer the research questions put forward in the Introduction, based upon both the theoretical insights and the conclusions drawn from the empirical chapters. In doing so, we summarise the key findings of the book and at the same time assess the (added) value of the policy arrangement approach as it has been developed thus far. Second, on that same basis of theoretical and empirical observations, we try to formulate a research agenda for the further development and refinement of the policy arrangement approach.

9.2 ANSWERS TO RESEARCH QUESTIONS

In the Introduction to this book five research questions were presented. Here we will ask them again and answer them one by one on the basis of both the theoretical and the empirical chapters. Where these answers are unsatisfactory or raise new questions, we will select issues for further deliberation. On the basis of that exercise, we then try to develop a research agenda in the next section.

1. What kind of environmental policy arrangements have been formed in environmental politics?

As has been revealed in chapters 1 to 3, the environmental policy domain proves indeed to be a theatre of increasing diversity and plurality in terms of predominant policy arrangements. What we and other academics commonly term as 'environmental policy' - thereby suggesting a certain homogeneity - actually encompasses a wide variety of topics, ranging from waste to landscape, from agriculture to mobility, and from nature conservation to climate. This variety of topics of course implies a variety of political problems and questions. In addition, these subdomains of environmental policy all have their own actors involved, their own set of coalitions and divides, their own acknowledged problem definitions, their own distribution of power and resources, etc. Hence these separate domains look for different ideological and institutional answers to their specific problems. This is reflected in a variety of policy arrangements, in the sense in which we theoretically defined this concept in the first three chapters.

The empirical chapters show and confirm the variety and plurality of policy arrangements which has emerged in the environmental field, both at national and international levels, although they by no means cover the full range of environmental policy domains. All chapters assert policy change and changing policy arrangements, but most also emphasise the existence and functioning, side-by-side, of different arrangements. Pestman (chapter 4) notes that there are two prevailing developments in Dutch infrastructure policies which look contradictory at first sight: on the one hand a development towards reflexive policy making with input from many stakeholders (*InfraLab*), on the other hand the (re)centralisation of decision making by the state (*Tracéwet*). This double-faced process has led to the side-by-side existence of different arrangements, both *interactive* and *statist*, in Dutch infrastructure policy, with both types of arrangement having its own rationale. Arts (chapter 6) concludes that in global environmental politics *interstatist*, *intergovernmental* and *transnational* arrangements exist side-by-side, although the first seems to become ever more obsolete, while the last is becoming more important. This is, among other reasons, due to the increasing role of non-governmental players in global environmental politics, in itself a 'container concept' for quite a variety of issues, political questions and policy arrangements. Wisserhof (chapter 8) shows that a proliferation of arrangements has taken place in the Dutch agricultural sector since the Second World War. In particular the breakdown of the so-called *Green Front*, traditionally characterised by its *corporatist* structure, and its development into a variety of new arrangements - *statist*, *civic* and *liberal* - have marked this occurrence. Again, the emergence and the institutionalisation of the environment as a key issue in agricultural policy is paralleled by an increasing variety of policy arrangements.

The other chapters lead to somewhat different conclusions with regard to the renewal and variety of policy arrangements, as they deal with domains in which a specific arrangement that is striven for has *not* so far emerged. De Jong (chapter 5) makes clear that in bilateral policy making on nature conservation the ascent of one *transboundary* arrangement which transcends those on both sides of the border is not obvious. A lack of similarity in national discourses and a low level of transboundary interaction have prevented the emergence of such an arrangement in the German-Dutch *Gelderse Poort*. The case illustrates the persistence of 'national styles' of policy making, including the national shading of certain arrangements. It gives rise to a number of questions about the various contingencies of a transnational, particularly a European, environmental policy. Janssens and Van Tatenhove (chapter 7) conclude that green planning in the Netherlands, despite its integrative ambitions, has not resulted thus far in *integrative* arrangements on physical planning, water management and the environment. There are, however, some promising developments, both at provincial level, where there are initiatives to integrate plans from the three policy domains, and at regional level, with experiments in collaborative planning. On the one hand the authors indicate that interdepartmental struggles over competencies and a general 'anti-planning mood' in the phase of late modernisation are among the main reasons for this 'failure of integration'. On the other hand they make clear that, over the years, the idea of environmental planning evolved from an ambitious, comprehensive 'blueprint' planning towards an 'integrative planning' meant to guide further negotiation, decision making and implementation. This process enabled the emergence of a range of policy experiments with a variety of more or less specific arrangements with regard to environmental, spatial and water management policies. But as for many of these projects implementation has started only recently, it is far too early to assess them properly.

These five chapters do not cover the full range of (innovative) policy arrangements in environmental politics. An important example not dealt with, but interesting from a policy arrangement perspective, is the series of arrangements which have emerged around so-called *covenants* in the Netherlands and, increasingly, in other European countries. Covenants are voluntary agreements between the state and market parties by which the latter freely adopt certain standards or targets, but under the guidance of - and sometimes under pressure from - the former. For example, in the *Benchmark* covenant that was recently adopted, the Dutch energy-intensive industry has promised to be at the top of the world's energy efficiency by the period 2008-2012, which is the target period of the Kyoto Protocol to the Climate Convention. In this, the industry is expected to contribute up to 10 per cent of the total Dutch commitment under this protocol (6 per cent less emission of greenhouse gases in the period 2008-2012 compared to 1990 levels). In turn, the Dutch government has promised not to impose any additional energy taxation upon this industrial sector. It would be worthwhile to analyse this example from a policy arrangement perspective, all the more so because neither the environmental movement nor any other civil society representative is part of this 'state-market' deal.

At the same time, we acknowledge the emergence of another type of 'covenant': a range of innovative arrangements of co-operation between civil society representatives and market agencies. Examples are initiatives by Greenpeace International and the Swiss firm Wenko AG to produce the *Smile* (a car using very little energy); by McDonalds, Conservation International and local NGOs to conserve nature in Central America; by Unilever and WWF to safeguard global fish stocks; by General Motors and the World Resources Institute to produce an extremely efficient engine; by WWF and the Body Shop to support sustainable development projects in the South; by Amnesty International, World Council of Churches and Shell to improve human rights and environmental conditions in developing countries, and others. Are these, as Beck suggests, the forerunners of an irreversible unravelling of centralised politics, which will include the emergence of differential partial arenas of cultural and social sub-politics, in which civil society representatives, clearly on a transnational basis, succeed in provoking private firms into investing in commonly designed projects? It would be interesting to investigate such cases from our perspective, as they are examples of what one could call 'policy arrangements without the state'. Even if most of them have hitherto only resulted in a kind of 'prototype' product or in rather experimental policies, they look promising initiatives for renewing environmental policy making in the near future. Yet one lacks a proper understanding of the (pre)conditions of their functioning and outcome.

The above makes clear that this book, given the cases we did and did not deal with, inadvertently may have slightly underestimated the role of the market. It seems that we, together with our colleagues in the environmental policy domain, emphasise the role of state agencies and civil society parties, and tend to neglect the role of market representatives and the new and specific kind of arrangements their entrance into the arena has led to. Due to processes of globalisation, privatisation and de- and re-regulation among others, business nowadays actually takes greater responsibility within environmental policy, or at least it claims and is asked to do so. Its role therefore should be given more prominence in our future research. This should include the actual functioning of different kinds of voluntary agreements or covenants with private agencies (with or without state agencies involved), not only of effectiveness, but also in terms of compliance, implementation, openness and legitimacy.

2. How do these environmental policy arrangements change?

In the theoretical chapters of the book we paid attention to at least four aspects of policy change. First, we identified two possible 'motors of change': political modernisation on the one hand and policy innovations on the ground on the other. Second, we distinguished between change in terms of policy content and change in terms of policy organisation.

Central to the policy arrangement approach is the notion of *duality of structure*. An important reason for developing the concept of policy arrangements is that it enables us to understand the institutionalisation of environmental policy making as the interplay between interactions and structural developments. Policy change induced by political modernisation provides an *institutional and structural* focus on social change, as a result of changing relations between state, civil society and market. Change stimulated by day-to-day interactions between actors is *strategic*, focusing on the arguments actors use in interactions, the norms and values they stand for and their problem definitions. The concept of policy arrangement should enable researchers to look at policy change from both perspectives.

We believe the empirical chapters, by using the concept of policy arrangement, indeed reflect this duality of structure, even if some chapters mainly emphasise political modernisation and institutional change. More specifically, several chapters analyse how policy arrangements in certain environmental domains have changed under the influence of different phases of political modernisation, characterised by new conceptions of governance and different state-market-civil society relations. Pestman (chapter 4) obviously does this for infrastructure policies, linking the policy change and the co-existence of apparently contradictory arrangements, in which the claims of state agencies and civil society representatives are balanced. At the same time he illustrates the strategic operations of both environmental and citizens' organisations. Wisserhof (chapter 8), covering different decades, looks at the changes in agricultural policies primarily within the context of changing state-market interrelations. Janssens and Van Tatenhove (chapter 7) clearly relate the developments in green planning to overall changes in the predominant style of governance and to the changing relations between state, civil society and market. At the same time, though, they emphasise the strategic role of spatial planning departments and the tension between experiments with interactive forms of planning and existing legislative and planning frameworks. Arts (chapter 6) begins by analysing international arrangements before he reconstructs the process of political modernisation and its impact upon global environmental politics. His analysis pays attention to institutional factors on the one hand and to the actor's perspective on the other, as he assesses the impact of new actors, e.g. global NGOs, entering the field. De Jong (chapter 5) takes a slightly different position, as her research is partly a reaction to the overestimating of institutional factors in the literature on transboundary policy making. As she looks primarily at the designing and evolution of one policy project, she more or less inevitably stresses the strategic behaviour of the parties involved, especially their (lack of) interactions, their (conflicting) perceptions and, in general, their discursive strategies.

In short, while the emphasis in individual chapters largely depends upon their empirical scope, we believe this book as a whole does justice to the 'duality of structure', in that it pays attention to structural and strategic factors equally. In so doing it takes a stance that is clearly distinguishable from some of the predominant approaches to environmental policy, such as the network approach, the configuration approach and the advocacy coalition framework. While the policy network and configuration approaches are mainly actor-oriented and voluntaristic in nature, and tend to neglect structural, institutional and power features of policy making, the advocacy coalition framework is more static and structuralistic. The policy arrangement approach bears most resemblance to the argumentative approach, although the focus is not so much on argumentative and discursive activity within a policy arrangement, but more on the institutionalisation of arrangements. With the 'political modernisation' thesis and the 'policy arrangement' concept we intended to provide a combination of strategic and structural approaches to policy making, by accentuating the institutionalisation of arrangements as being the interplay between discursive activity and processes of social and political change.

In the theoretical chapters of this book we went one step further in the elaboration of our institutional and strategic perspectives, by introducing an *organisational* and a *substantive* focus to analyse social change. Our definition of a policy arrangement presumes a well-balanced attention to both, as we conceive of the dynamics and dialectics of changes in substance and changes in organisation as the quintessence of policy change. In chapter 3 we operationalised these two aspects into four dimensions: 'policy coalitions', 'resources and power', 'rules of the game' and 'policy discourse'.

Most empirical chapters take either organisation or substance as their main starting point, without however neglecting the other side of the coin. The analyses by Pestman (chapter 4), De Jong (chapter 5), and Janssens and Van Tatenhove (chapter 7) are more substantive than organisational in nature, as they pay a great deal of attention to (changing) *discourses* on infrastructure, nature policy and integrated planning respectively. They all perceive discourse, or more precisely discursive innovations, as an important vehicle for the changing of policy arrangements, whereas a stable (or less changeable) discourse might hamper such change. The 'mainport' discourse, for example, has prompted the design of new infrastructural mega-projects in the Netherlands, leading to and obviously legitimising new, centralised forms of policy making. The persistence of different discourses on nature and nature conservation on both sides of the Dutch-German border - and the lack of interactions between the advocates of these discourses - has prevented the emergence of a common, transboundary policy arrangement. The other chapters have a different starting point and are more organisational than substantive in nature. Arts (chapter 6) emphasises the entrance into the arena of new players and the emergence of new coalitions in global environmental policy making as the instigators of change *par excellence*, as they lead to new power relationships and to new rules of the game. Wisserhof (chapter 8) analyses the breakdown of the corporatist structure that formerly existed in Dutch agricultural policies primarily in terms of changing rules and resources, which he believes to be the main factors explaining the policy changes.

It follows from the above that both discursive changes and the appearance of new actors are studied in some detail as major instigators of change, whereas the analysis of power and rules - as vehicles for or obstacles to change - remains less developed. This is quite evidently a result of the fact that both concepts are difficult to operationalise. This is particularly so for 'power', an essentially contested concept. Some students in our research programme already have taken new initiatives in order to elaborate on power in policy arrangements, but these efforts should be doubled in future research (see below).

Nevertheless, we believe the perspective developed and detailed in this book, especially the policy arrangements concept, will prove to be profitable in analysing policy change, in particular because it pays attention to discursive and organisational factors, as shown in coalitions, rules, resources and discourses. We cannot and will not grade these different motors of change, let alone indicate one of them as of fundamental importance, because we believe that the specific (added) value of our approach lies precisely in its encompassing nature, paying attention to the dynamics of institutional and strategic, discursive and organisational change. We neither theoretically nor empirically believe that one can identify a single factor as the key to explain policy change (or the lack thereof). In this sense we can best typify change by using the metaphor of a *lemniscate*. In all empirical cases we see an avalanche of changes. Following the loops of the *lemniscate* these changes have sometimes been initiated by new coalitions, in other cases provoked by innovative discourses, have been reinforced by rules and resources, and have set off a chain reaction of changes in all aspects. Whether institutional or strategic, organisational or substantive factors initiate the domino effect depends upon the empirical case under investigation. This overall conclusion on the dynamics and dialectics of policy change raises new questions, such as under which circumstances and through which mechanisms a chain reaction occurs or is stopped. Comparative research, either in different domains or in different countries, might open new perspectives upon these and related questions (see next section on the research programme).

3. Is there a shift from one type of environmental policy arrangements to other types? If so, how can these shifts be explained?

This third question refers to what is called the *diachronic* analysis in chapter 3, to the change in policy arrangements over time and to probable common trends in that change. All chapters use such a diachronic perspective, but at the same time they all make clear that in the process of political modernisation there is no unilinear evolutionary shift from one type of policy arrangement to another. On the contrary, because as mentioned earlier, both Pestman (chapter 4) and Wisserhof (chapter 8) indicate 'two faces of political modernisation': one leading to an increasing plurality of more or less interactive or reflexive arrangements, the other to a (re)centralisation of state power. In addition, both Pestman and Wisserhof refer to the current juxtaposition of different arrangements, some of which can be characterised as more *innovative* in nature (or 'late-modernist'), others as *traditional* (or 'early-modernist'). In a similar way, Arts notes that in certain issue-areas of global environmental policy making - such as forests - interstatist relationships are still dominant, despite innovative developments in other areas. Some other chapters show, as we have seen, that an expected shift to a particular arrangement is not taking place, while almost all chapters establish the existence and persistence of different, diverging or even contradictory arrangements side-by-side.

These observations clearly correct the widespread assumption, discussed in chapter 1, of a unilinear modernisation of environmental politics, something which underestimates the dynamics and plurality of the process. As the empirical chapters show, there is no clear and predominant movement away from (inter)statist arrangements, in which the state plays the dominant role, towards types of arrangements in which ever more non-governmental and private stakeholders and coalitions have become legitimate and authoritative players. These latter arrangements are referred to as 'civic', 'plural', 'liberal', 'transnational' and 'interactive' in the literature and in the various chapters of this book. Instead of a unilinear trend, we see the state withdrawing from some areas while increasing its influence in others. We see new arrangements between state and civil society being established, while elsewhere the state adheres strictly to its privileges. We see patterns of traditionally privileged interaction between state and market being broken down, while such patterns are re-established in other domains. These conclusions also apply to domains which have not been dealt with extensively in this book, but in which we have been engaged in previously, such as waste and energy policies, transport and traffic policies, environmental business management and others.

Given this plurality of co-existing arrangements, which vary among other things with regard to the state's role, one should definitely put the 'withdrawing of the state' thesis and, even more, the 'state power erosion' thesis in perspective. It seems more useful to think of the state as taking varying roles in the process of political modernisation, differing also from domain to domain and, of course, from country to country. Without suggesting a unilinear shift, a variety of state roles is possible. One can think of a 'providing' or an 'enabling' state (in domains such as energy, water, waste and similar policies), of a state taking a 'director's role' or a 'negotiator's role' and of a 'regulating' or a 'transactional' state, in its interaction with both market and civil society. In order to do justice to the actual plurality and variety of environmental policy arrangements, these and other dichotomies and distinctions should not be regarded as more or less obligatory *points de passage* in a predetermined process of political change, but as analytically useful ideal-types. The overall conclusion should be that political modernisation does not imply a coherent path with regard to the development of environmental policy arrangements, but that it exhibits several faces, divergent directions and a variety of institutional outcomes.

These conclusions lead to more or less self-evident questions for further research, dealing with the why and how of different and even contradictory policy changes in various domains, in different countries, and the like. The assumption and acceptance of this plurality and variety of arrangements also raises the question of their *categorisation.* So far we have distinguished different arrangements, without providing coherent and consistent labelling. One may instinctively know what is meant by a 'statist', 'corporatist' or 'liberal' arrangement, and to what extent they differ, but a consistent categorisation of 'ideal-types' of arrangements is desirable.

4. Do different types of environmental policy arrangements co-exist in certain periods of environmental politics? If this is the case, how can this co-existence be explained?

This question refers to what is called the *synchronic* analysis in chapter 3. To a large extent it has already been answered above, since several chapters definitely conclude that there is a co-existence of different types of policy arrangements in many policy domains, in *any* phase of political modernisation. Yet specific arrangements, such as the statist one in the phase of early modernisation, are clearly predominant in any given era.

Although this co-existence is well described in several chapters, the authors are less successful in explaining the phenomenon, as most provide classical, partial or specific explanations. De Jong's explanation (chapter 5) is a classical one, in that she makes clear that traditional, nationally-diverging discourses on the one hand and deeply-rooted national styles of policy making on the other explain the existence and persistence of two country-specific types of policy arrangements. One transboundary project, however ambitious and prestigious, apparently does not suffice to overrule these differences. This raises questions about the need and the opportunities for a more convergent (cross-national) policy making, for instance when implementing international or supranational policies. Janssens and Van Tatenhove (chapter 7) come to a similarly classical conclusion when they argue that domain-specific systems and styles of planning impede the integration of these policy domains. Wisserhof (chapter 8) puts forward a shock event (the environmental crisis in agriculture) and the conservative attitude of some players (farmers, members of parliament, the ministry) to explain the co-existence of both innovative and traditional policy arrangements in agriculture. The former explanation appears quite specific at first sight. And yet it raises the question as to why the environmental issue and rationale is perceived and handled as a crisis in some domains - appealing for and leading to new arrangements - whereas other domains and their pre-existing arrangements seem capable of dealing with these new policy challenges more easily. Pestman (chapter 4) suggests that the renewal of arrangements in infrastructure policies was the result of public pressure and the need for legitimacy (in terms of public support) for these policies. The persistence and the recent reinforcement of statist policy making in this domain, on the other hand, is caused by pressure from within the political sphere itself to speed up decision making, which can be called legitimacy in terms of the capabilities of the political system. Hence, the pressure from different societal subsystems, e.g. civil society versus state, each of them translated into different discourses and giving rise to different coalitions of actors, has resulted in the emergence of various policy arrangements in one domain and in one single period. Arts (chapter 6), finally, might be the most promising in terms of explaining the persistence of traditional policy arrangements, as he links this phenomenon to the kind of conflict which the policy issue concerned occasions. He suggests that confrontation and a lack of consensus tend to lead to interstatist ways of dealing with an issue (e.g. the forest case), whereas integration and consensus tend to produce transnational policy arrangements (e.g. the biodiversity case). His thesis seems promising, because he links the most probable kind of arrangement directly to the type of issue involved and to the political coalitions and divisions it gives rise to. At the same time, though, his suggestions raise questions about who is capable of framing an issue in a specific way, thereby deliberately or unintentionally influencing its most probable type of institutional arrangement.

In short, different national discourses, departmental struggles, shock events, the attitudes of players, diverging public and political pressures and the 'naming' and 'framing' of the issues may all be factors which explain the co-existence of several policy arrangements in any one domain and in any one period. But this list of variables, drawn from the empirical chapters of this book, is obviously incomplete and, both theoretically and analytically, unsatisfactory in understanding and explaining this co-existence of plurality and variety. We therefore add the design of a better, more comprehensive and more consistent explanatory scheme to the priorities for further research.

5. To what extent are policy domains organised into specific arrangements? In other words, are environmental policy arrangements issue-specific?

From the above it should be clear that this final research question cannot be answered by an unequivocal 'yes' or 'no'. On the one hand, policy arrangements do not seem to be issue-specific, as for example Pestman (chapter 4) and Wisserhof (chapter 8) identify both statist and more interactive arrangements in infrastructure and agricultural policies respectively. De Jong (chapter 5) makes clear that the organisation of a single policy domain, e.g. nature conservation, in two separate countries has led to fundamentally different arrangements. On the other hand, though, Janssens and Van Tatenhove (chapter 7) emphasise that the domain-specific characteristics of policy planning arrangements are an important obstacle to their integration, while Wisserhof points to the apparently 'unique' linkage between the agricultural sector or agricultural issues and a corporatist arrangement. One should realise however that this type of arrangement is much more widespread. We can think here of tripartite structures in social and industrial policy making in Western nation states and, for instance, in the International Labour Organisation. As stated earlier, we might even see the (re)establishment of some corporatist-style arrangements in environmental policies at European and national level in respect of industries which have recently developed. Arts (chapter 6) notes that in global environmental policies different issues tend to produce different arrangements. But he makes it very clear that it is not the issue itself which is important, but the nature and characteristics of the social and political conflicts which it provokes. The kind of actors involved in an issue, the coalitions and divides it causes, and the pre-existing and well-established political structures it is part of are the determinant characteristics. It is therefore perfectly possible, for example, that a more transnational arrangement will emerge in the forest domain once the North-South conflict on this issue is resolved. In other words, it is not the forest issue itself which tends to produce interstatist policy making, but the fact that this issue runs parallel with a deep-rooted political division.

Again, our analyses raise new questions. First of all we might look for typical, *level-* rather than issue-specific characteristics in arrangements. Chapters 5 and 6 rightly state that arrangements at bilateral and global level are of a different nature to those at national level. After all, arrangements at levels above that of the nation state lack a central and authoritative sovereign power, and therefore are largely based on consensus decision making, relying on voluntary compliance and without the power to enforce such decisions, as both De Jong and Arts emphasise in their chapters. The EU, with some supranational characteristics, is to some extent an exception to these rules, but EU policies have remained outside the scope of this book. In future research, these principal differences between the international and the national level should be studied carefully by examining arrangements which have recently become established and are now developing. Even within the EU one sees various institutional arrangements in an attempt to overcome the problem of sovereignty. The same applies, *a fortiori*, to international arrangements. It seems to us worthwhile to compare them systematically from this perspective.

Secondly, one could argue that arrangements at different levels are highly interdependent and therefore have an impact on each other. One instinctively feels that, for instance, a changing arrangement in waste management at the national level due to increasing private-public co-operation, changes the degree of freedom local authorities have in their waste policies. The same applies of course to the interaction between national and supra- or international arrangements. On climate change, for example, prominent discourses and policy standards which have been adopted in the Netherlands are to a large extent based on international debates and decisions, both within the UN and the EU. Consequently, one should not just study arrangements at different levels, but include their interrelations and patterns of influence as well. We believe this issue of the interdependence of arrangements at different levels of policy making has so far not been sufficiently addressed.

9.3 A FUTURE RESEARCH AGENDA

In answering the five research questions, a number of issues for further deliberation, discussion and research have been identified. These form the basis on which we propose a specific research agenda for the near future. In order to categorise and prioritise the questions raised above, we perceive three starting points for such a research agenda: theoretical, empirical, and political.

Theoretical questions

While we believe that the approach developed thus far has proved fruitful, the basic concepts still need further elaboration and balancing. This applies, first of all, to the combined analysis of the institutionalisation of policy arrangements from an institutional perspective and from a perspective of strategic conduct. Secondly, a further balancing of the organisational and substantive focus, and its development in the dimensions 'policy coalitions', 'resources and power', 'rules of the game' and 'policy discourse', is needed in order to understand the dynamics and dialectics of change.

It might be useful, in order to fully acknowledge the duality of structure, to set up a double research project, in which one researcher deliberately starts from an institutional and the other from a strategic perspective, while examining the same domain. Equally, because we believe that the organisational perspective is less detailed particularly as far as the analysis of the rules of the game and power is concerned, research focusing on these aspects should be initiated. So far, students in our research programme have built on (a) rule interpretations from various sources, e.g. arena models and regulation theories, and (b) reputation and resource mobilisation methods, to operationalise the concept of power. The question remains, however, to what extent these approaches, models and methods match each other and our theoretical premises. More theoretical and methodological refinement on these issues is definitely needed.

Secondly, we believe our approach to and theory of the diachronic development of policy arrangements is reasonably fully expanded, with concepts such as 'political modernisation', 'policy change' and 'policy innovation' (although the last is less elaborated upon). Our theoretical insights into the synchronic development of policy arrangements, however, are not up to that level. Although individual authors refer to several variables which might explain the co-existence of traditional and innovative arrangements, we still need to develop a consistent explanatory theory on this issue. One might consider a theory on the dynamics and dialectics of change, explaining the persistence of certain arrangements and the coming into being of new ones in the same domain at the same time. In terms of our approach and concepts, the following variables could be thought of as relevant: (a) the (countervailing) power of (coalitions and oppositions between) actors to change or conserve policy processes; (b) the enabling or constraining effects of the rules of the game on the degree of freedom players have to persist with or change their behaviour; and (c) the 'clash' of old and new policy discourses and their ability to mobilise new or, on the contrary, preserve existing arrangements. A systematic elaboration, both theoretically and empirically, of these variables would definitely enrich our approach.

Thirdly, as indicated above, there is a need for a consistent categorisation of ideal-types of policy arrangements. The labels 'statist', '(neo-)'corporatist' and 'civic' may have been adequate hitherto - although they themselves all gave rise to scientific controversies - but they no longer suffice for the variety and plurality of arrangements referred to in this book and elsewhere. We still believe such a categorisation should be based primarily upon the following questions: (a) which of the three subsystems (state, market, and civil society) is involved? (b) how are their interrelations organised? (c) which of the subsystems is predominant in a certain arrangement? and (d) what substantive effects do specific organisational arrangements have and *vice versa*? In addition, we should clearly distinguish between arrangements in international relations on the one hand and those within national political systems on the other.

Empirical questions

It will be clear that answering the theoretical questions above in itself asks for further empirical research. Here, however, we want to deal with some empirical questions which have so far tended to be neglected in our research and in this book.

First, as the book does of course not cover all innovative developments in environmental policy making, new cases should be studied to enrich the policy arrangement approach. One could think here of policy domains which have not been covered, some of which were mentioned earlier (such as energy, waste, traffic and business management). But, starting from the policy arrangement perspective, we are not thinking so much about other domains as about other types of arrangements. As stated earlier, thus far we have tended to ignore those arrangements, often labelled 'private-public partnerships', 'voluntary agreements', or 'covenants', in which market agencies play a crucial role, whether as partner of the state, or of civil society, or of both. One area we already study is climate change. Particularly in this issue-area, the role of business has changed dramatically since the adoption of the Kyoto Protocol in 1997: in general one sees the replacement of a defensive strategy by industry with a proactive one. As a consequence, a series of new policy arrangements is now under construction in this field, at different policy levels and between states and companies as well as between NGOs and companies. At least in Western countries, these arrangements are constructed within the context of a rapidly changing organisation of energy policies, due to various forms of privatisation. Both the new arrangements in this field and newly developing ones in other domains formerly organised as public utilities, such as waste, water supply, public transport and others, deserve further analysis.

Secondly, as mentioned above, we most certainly have to pay more attention to the interaction of arrangements at different political levels. Due to the increasing Europeanisation of politics ever more policy initiatives stem from Brussels, and therefore more emphasis should be placed on EU policies, their impact on national policy arrangements and, *vice versa*, the impact of (differing) national arrangements on EU policies. Currently, for example, some 50 per cent of new Dutch environmental legislation has its origin in EU initiatives. We have already started a PhD project on the basis of such a multi-level approach, which aims to analyse both European and national nature conservation and nature development policies. The project examines the changing character of arrangements at EU level in terms of actors, discourses, rules and resources. Recent EU policy plans, directives and funds reflect a major change in such arrangements and in the division of roles and tasks between the EU, the nation states and the regions within the member states. At the same time these changes run parallel with rapidly changing arrangements in this field at regional and local levels within the member states, which in their turn influence European regulations.

Political questions

Finally, our research programme inevitably deals with political questions which we have to resolve. We confine ourselves here to clearly marking out our position in terms of description and prescription, in other words, of analysis and intervention. In the first chapters we clearly distinguished our position from that of many colleagues, who consider the design and improvement of policy strategies to be an intrinsic, if not the primary, task of a political scientist. In this book we developed the policy arrangement concept primarily as an analytical concept, and not as a prescriptive one, let alone as an interventionist tool. Central in the policy arrangement approach as an *analytical* model is, we hope, its well-balanced attention to stability and change, to tradition and renewal, and to the interplay between structure and actor. It is only from this analytical model that it is possible to develop the policy arrangement approach into an *interventionist* model. Such an interventionist model would identify those variables which could be helpful to policy makers in 'arranging' a policy field, for example in order to increase the effectiveness and legitimacy of policy measurements. The interventionist model also needs to be based on the notion of duality of structure and on an analysis of the four dimensions: coalitions, rules of the game, resources and policy discourses.

Just like politicians and civil servants, we are interested in the circumstances and conditions of policy change, or more precisely, in the how and why of the shifting from one arrangement to another, or the absence of such a shift. However, application of the interventionist model emphatically has to be based on the findings of the policy arrangement approach as an analytical model. This prevents us from looking at events solely from an actor's perspective, while also not - at least so we hope - emphasising unilaterally the strategic opportunities for policy change. Without having the ambition of steering the renewal of environmental policy arrangements, it is therefore absolutely necessary to study under which conditions policy arrangements can or cannot change and, in the final analysis, can *be* changed. We believe our best contribution to these knowledge interests is not to pretend that we could ever manage arrangements freely and form and change them at will. On the contrary, it seems to us that further theoretical and empirical research will enable us to identify the most important conditions and variables, while at the same time putting the rationality claims of policy making critically into perspective.

INDEX

U

ENVIRONMENT & POLICY

1. Dutch Committee for Long-Term Environmental Policy: *The Environment: Towards a Sustainable Future*. 1994 ISBN 0-7923-2655-5; Pb 0-7923-2656-3
2. O. Kuik, P. Peters and N. Schrijver (eds.): *Joint Implementation to Curb Climate Change*. Legal and Economic Aspects. 1994 ISBN 0-7923-2825-6
3. C.J. Jepma (ed.): *The Feasibility of Joint Implementation*. 1995 ISBN 0-7923-3426-4
4. F.J. Dietz, H.R.J. Vollebergh and J.L. de Vries (eds.): *Environment, Incentives and the Common Market*. 1995 ISBN 0-7923-3602-X
5. J.F.Th. Schoute, P.A. Finke, F.R. Veeneklaas and H.P. Wolfert (eds.): *Scenario Studies for the Rural Environment*. 1995 ISBN 0-7923-3748-4
6. R.E. Munn, J.W.M. la Rivière and N. van Lookeren Campagne: *Policy Making in an Era of Global Environmental Change*. 1996 ISBN 0-7923-3872-3
7. F. Oosterhuis, F. Rubik and G. Scholl: *Product Policy in Europe: New Environmental Perspectives*. 1996 ISBN 0-7923-4078-7
8. J. Gupta: *The Climate Change Convention and Developing Countries: From Conflict to Consensus?* 1997 ISBN 0-7923-4577-0
9. M. Rolén, H. Sjöberg and U. Svedin (eds.): *International Governance on Environmental Issues*. 1997 ISBN 0-7923-4701-3
10. M.A. Ridley: *Lowering the Cost of Emission Reduction: Joint Implementation in the Framework Convention on Climate Change*. 1998 ISBN 0-7923-4914-8
11. G.J.I. Schrama (ed.): *Drinking Water Supply and Agricultural Pollution*. Preventive Action by the Water Supply Sector in the European Union and the United States. 1998 ISBN 0-7923-5104-5
12. P. Glasbergen: *Co-operative Environmental Governance: Public-Private Agreements as a Policy Strategy*. 1998 ISBN 0-7923-5148-7; Pb 0-7923-5149-5
13. P. Vellinga, F. Berkhout and J. Gupta (eds.): *Managing a Material World*. Perspectives in Industrial Ecology. 1998 ISBN 0-7923-5153-3; Pb 0-7923-5206-8
14. F.H.J.M. Coenen, D. Huitema and L.J. O'Toole, Jr. (eds.): *Participation and the Quality of Environmental Decision Making*. 1998 ISBN 0-7923-5264-5
15. D.M. Pugh and J.V. Tarazona (eds.): *Regulation for Chemical Safety in Europe: Analysis, Comment and Criticism*. 1998 ISBN 0-7923-5269-6
16. W. Østreng (ed.): *National Security and International Environmental Cooperation in the Arctic – the Case of the Northern Sea Route*. 1999 ISBN 0-7923-5528-8
17. S.V. Meijerink: *Conflict and Cooperation on the Scheldt River Basin*. A Case Study of Decision Making on International Scheldt Issues between 1967 and 1997. 1999 ISBN 0-7923-5650-0
18. M.A. Mohamed Salih: *Environmental Politics and Liberation in Contemporary Africa*. 1999 ISBN 0-7923-5650-0
19. C.J. Jepma and W. van der Gaast (eds.): *On the Compatibility of Flexible Instruments*. 1999 ISBN 0-7923-5728-0
20. M. Andersson: *Change and Continuity in Poland's Environmental Policy*. 1999 ISBN 0-7923-6051-6

ENVIRONMENT & POLICY

21. W. Kägi: *Economics of Climate Change: The Contribution of Forestry Projects*. 2000
ISBN 0-7923-6103-2
22. E. van der Voet, J.B. Guinée and H.A.U. de Haes (eds.): *Heavy Metals: A Problem Solved?* Methods and Models to Evaluate Policy Strategies for Heavy Metals. 2000
ISBN 0-7923-6192-X
23. J. van Tatenhove, B. Arts and P. Leroy (eds.): *Political Modernisation and the Environments*. The Renewal of Environmental Policy Arrangements. 2000
ISBN 0-7923-6312-4

KLUWER ACADEMIC PUBLISHERS – DORDRECHT / BOSTON / LONDON

MIX
Papier aus verantwortungsvollen Quellen
Paper from responsible sources
FSC® C105338

If you have any concerns about our products,
you can contact us on
ProductSafety@springernature.com

In case Publisher is established outside the EU,
the EU authorized representative is:
Springer Nature Customer Service Center GmbH
Europaplatz 3, 69115 Heidelberg, Germany

Printed by Libri Plureos GmbH
in Hamburg, Germany